INDUSTRY STANDARD
OF THE PEOPLE'S REPUBLIC OF CHINA

Code for Design of Railway Passenger Station

TB 10100-2018

Prepared by: China Railway Design Corporation
Approved by: National Railway Administration
Effective date: September 1, 2018

China Railway Publishing House Co.,Ltd.

Beijing 2020

图书在版编目(CIP)数据

铁路旅客车站设计规范:TB 10100-2018:英文/中华人民共和国国家铁路局组织编译.—北京:中国铁道出版社有限公司,2020.6
ISBN 978-7-113-25604-3

Ⅰ.①铁… Ⅱ.①中… Ⅲ.①铁路车站-客运站-建筑设计-设计规范-中国-英文 Ⅳ.①TU248.1-65

中国版本图书馆 CIP 数据核字(2019)第 042606 号

Chinese version first published in the People's Republic of China in 2018
English version first published in the People's Republic of China in 2020
by China Railway Publishing House Co., Ltd.
No. 8, You'anmen West Street, Xicheng District
Beijing, 100054
www.tdpress.com

Printed in China by Beijing Hucais Culture Communication Co., Ltd.

ISBN 978-7-113-25604-3

About the English Version

The translation of this Code was made according to Railway Engineering and Construction Development Plan of the Year 2016 (Document GTKFH [2016] 29) issued by National Railway Administration for the purpose of promoting railway technological exchange and cooperation between China and the rest of the world.

This is the official English language version of TB 10100-2018. In case of discrepancies between the original Chinese version and the English translation, the Chinese version shall prevail.

Planning and Standard Research Institute of National Railway Administration is in charge of the management of the English translation of railway industry standard, and China Railway Economic and Planning Research Institute Co. , Ltd. undertakes the translation work. Beijing Times Grand Languages International Translation and Interpretation Co., Ltd. provided great support during translation and review of this English version.

Your comments are invited and should be addressed to China Railway Economic and Planning Research Institute Co., Ltd., 29B, Beifengwo Road, Haidian District, Beijing, 100038 and Planning and Standard Research Institute of National Railway Administration, Building B, No. 1 Guanglian Road, Xicheng District, Beijing, 100055.

Email: jishubiaozhunsuo@126. com

The translation was performed by Yang Sibo, Dong Suge, Chai Guanhua, Dong Xuewu, Zhang Lixin, Wang Xueyuan.

The translation was reviewed by Chen Shibai, Wang Lei, Yang Quanliang, Jiang Jinhui, Yuan Li, Zhang Jialu.

Notice of National Railway Administration on Issuing the English Version of 18 Railway Standards including *Code for Design of Railway Passenger Station and a Series of Standard for Acceptance of Railway Works*

Document GTKF [2019] No. 45

The English version of *Code for Design of Railway Passenger Station* (TB 10100-2018) and a series of standard for acceptance of railway works is hereby issued (see the table below). In case of discrepancies between the Chinese version and the English version, the former shall prevail.

China Railway Publishing House is authorized to publish the English version of these standards.

List of the English Version of 18 Standards

S/N	Title	Reference No.
1	*Code for Design of Railway Passenger Station*	TB 10100-2018
2	*Standard for Acceptance of Earthworks in Railway*	TB 10414-2018
3	*Standard for Acceptance of Bridge and Culvert Works in Railway*	TB 10415-2018
4	*Standard for Acceptance of Tunnel Works in Railway*	TB 10417-2018
5	*Standard for Acceptance of Track Works in Railway*	TB 10413-2018
6	*Standard for Acceptance of Communication System in Railway*	TB 10418-2018
7	*Standard for Acceptance of Signaling System in Railway*	TB 10419-2018
8	*Standard for Acceptance of Electric Power System in Railway*	TB 10420-2018
9	*Standard for Acceptance of Traction Power Supply System in Railway*	TB 10421-2018
10	*Standard for Acceptance of Earthworks in High-speed Railway*	TB 10751-2018
11	*Standard for Acceptance of Bridge and Culvert Works in High-speed Railway*	TB 10752-2018
12	*Standard for Acceptance of Tunnel Works in High-speed Railway*	TB 10753-2018
13	*Standard for Acceptance of Track Works in High-speed Railway*	TB 10754-2018
14	*Standard for Acceptance of Communication System in High-speed Railway*	TB 10755-2018
15	*Standard for Acceptance of Signaling System in High-speed Railway*	TB 10756-2018
16	*Standard for Acceptance of Electric Power System in High-speed Railway*	TB 10757-2018
17	*Standard for Acceptance of Traction Power Supply System in High-speed Railway*	TB 10758-2018
18	*Standard for Acceptance of Concrete Works in Railway*	TB 10424-2018

National Railway Administration

December 24, 2019

Notice of National Railway Administration on Issuing Railway Industry Standard (Engineering and Construction Standard Batch No. 3, 2018)

Document GTKF [2018] No. 54

Code for Design of Railway Passenger Station (TB 10100-2018) is hereby issued and will come into effect on September 1, 2018. *Code for Design on Accessibility of Railway Passenger Station Buildings* (TB 10083-2005) is withdrawn.

National Railway Administration
June 11, 2018

Foreword

Railway passenger station is an important part of the railway network, an important node in the integrated traffic system and a crucial engine for the regional economic development. Since the issuance of *Code for Design of Railway Passenger Station Buildings* (GB 50226-2007), more than 1 000 new railway passenger stations have been built in China, including more than 300 medium or larger passenger stations. These new stations, by implementing the guidelines of "innovation, coordination, environmental protection, openness and sharing", giving overall consideration to the requirements of railway transportation, integrated transportation system development and urban development, and striving for the harmonization between transportation building, spirit of the time, and local characteristics, have surpassed the previous stations in terms of design concept, functional configuration, architectural style, technological innovation, effects of operation, etc. The large-scale integrated transportation hubs such as Beijing South Railway Station, Shanghai Hongqiao Railway Station, Guangzhou South Railway Station and Wuhan Railway Station, as well as those railway passenger stations with salient local features such as Lhasa Railway Station, Yan'an Railway Station, Suzhou Railway Station and Sanya Railway Station, have become the representative works for the construction of railway passenger stations, offering rich experience for further refining the standards of railway passenger stations.

The revision is conducted by fully summarizing the practical experience and research findings concerning the construction and operation of railway passenger stations in China. In this revision, the general policy on coordinating economic, political, cultural, social, and ecological aspects is implemented, the requirements on building modern integrated transportation hub and on "zero distance" transfer are considered with the purpose of optimizing the functional configuration of railway passenger station and improving the quality and efficiency of traveling; the safety requirements for cross-track facilities, structural design, inspection and maintenance, etc. are added in line with the principle of prioritizing safety, with the purpose of guaranteeing passenger safety; the national policy of environmental protection is implemented with the purpose of providing technical support for building environmentally friendly railway passenger stations; the technical parameters are optimized in order to further improve the technical and economic rationality; the scope of application is enlarged to render this Code more comprehensive; the name of this Code is revised as *Code for Design of Railway Passenger Station*.

This Code consists of 12 chapters, namely General Provisions, Terms, Overall Design, General Plan, Station Building, Facilities for Passenger Transport, Structure, Heating, Ventilation and Air Conditioning, Water Supply and Drainage, Electric Power and Lighting, Passenger Transport Service Information System, and Accessible Facilities.

The main revisions are as follows:

1. The concept of modern integrated transportation hub is materialized, and the requirement on "zero distance" transfer is put forward.

2. The chapter "Overall Design" is added to strengthen the systematic design of railway passenger station.

3. The requirements for green building design are added.

4. The original two chapters "Location and Overall Layout" and "Station Square" are combined into one chapter "General Plan", in which the requirements for the supporting urban traffic facilities are supplemented.

5. The requirements on the area for security check and identity verification as well as the requirements on transfer flow line are added.

6. The scale indexes for concourse and waiting area (hall/room) are revised, the requirements on commercial areas inside stations and on the total construction area of station building are added.

7. The design requirements for inspection and maintenance facilities are added.

8. The following contents are added: underground station, indoor environment, decoration and details, curtain wall and metal roofing, building energy conservation, lift and escalator, passenger guidance system.

9. The original chapter "Firefighting and Evacuation" is deleted.

10. The chapter "Structure" is added, which puts forward the requirements on special structures such as "bridge-building integrated" structure and long-span roof.

11. The original chapter "Building Equipment" is divided into four chapters, namely "Heating, Ventilation and Air Conditioning", "Water Supply and Drainage", "Electric Power and Lighting", and "Passenger Transport Service Information System", with some contents being revised.

12. The chapter "Accessible Facilities" is added.

The Technology and Legislation Department of National Railway Administration is responsible for the interpretation of this Code.

We would be grateful if anyone finding the inaccuracy or ambiguity while using this Code would inform us and address the comments to China Railway Design Corporation (No. 10, Zhongshan Road, Hebei District, Tianjin, 300142) and China Railway Economic and Planning Research Institute Co.,Ltd. (No. 29B Beifengwo Road, Haidian District, Beijing, 100038) for the reference of future revisions.

Chief Technical Leaders:

Zheng Jian, An Guodong, Wu Kefei

Prepared by:

China Railway Design Corporation

Drafted by:

Zhou Tiezheng, Zhao Jianhua, Du Shuang, Liu Yan, Lu Ying, Hou Kai, Yu Yang, Xu Xianghui, Li Guiping, Lv Xiaozheng, Xia Tianyan, Zhao Shilei, Sun Hongfeng, Meng Ran, Jie Beilei, Zhang Jialu, Xing Jiayong, Yu Fengfeng, Shi Yuezhen, Tang Hu, Zhang Jianxin, Zou Zhisheng, Dong Cheng, Zhu Jianzhang, Wen Jing, Tao Ran, Feng Shijie, Fan Yan, Hong Wei, Li Ling, Ma Jingbo, Feng Jingran, Zhao Chang, Dong Zhiqiang, Fang Jian, Guo Ruixia, He Tao, Wang Li, Li Maosheng, Li Zheng, Shen Yuerong, Huang Baomin, Guo Weidong, Ma Minjie, Li Guofu, Lv Yina, Yang Weihong, Mou Zhongxia.

Reviewed by:

Wu Kefei, Han Zhiwei, Zeng Huixin, Yu Zufa, Sang Cuijiang, Li Jing, Sheng Hui, Jin Xuwei, Li Chunfang, Guo Ming, Yu Shiping, Jing Deyan, Ye Nianfa, Qiu Jian, Chen Qiang, Li Yingjiu, Wang Zhehao, Lian Wenbin, Ning Fei, Wang Min, Chen Jun, Yang Sibo, Tian Yang, Tan Yueren, Jiang Wenxing, Liu Xun, Fan Liuchang, Zhang Min, Wang Yu, Han Guoxing, Hu Xiaoyong, Wei Wei, Wang Mu, Zhang Yali, Wang Xiangdong, Ye Di, Liu Jieping, Gong Yun, Yan Wei, Guo Xuhui, Guan Yamin, Chen Qihui, Song Ge, Xue Feng, Zheng Duo, Rong Huaqiang, Gan Bojie, Ma Zhijun, Shan Songtang, Hou Bin, Zhang Keyi, Li Mingguo, Bi Qinghuan, Jiang Jinhui, Li Zhengtao, Liu Hui.

Record of updates:

Code for Design of Railway Passenger Station Buildings (GB 50226-95), GB 50226-2007, GB 50226-2007 (2011 Version).

Contents

1 General Provisions

1.0.1 This Code is formulated with a view to meeting the demand of railway development in China, unifying the design standards of railway passenger stations, making the design of railway passenger stations comply with the requirements for safety, reliability, technological soundness, convenience, comfort, energy conservation, environmental protection, economy, and applicability.

1.0.2 This Code is applicable to the design of new and upgraded railway passenger stations.

1.0.3 Railway passenger stations shall have reasonable scale, complete functions, smooth flow lines and proper standards, in order to provide a convenient and comfortable boarding-alighting environment for passengers.

1.0.4 Effective measures shall be taken to guarantee the safety of passengers and their belongings, considering the high density of people and the great impact of stations on safety of train operation.

1.0.5 The design of railway passenger station shall comply with the requirements of overall urban planning and integrated transportation system planning, and shall be in harmony with the cultural and natural aspects of the city.

1.0.6 The design of railway passenger station shall embody the characteristics of transportation buildings and reflect the local features, traditional culture and modern spirits.

1.0.7 The design of railway passenger station shall comply with the requirements on modern integrated transportation hub and on "zero distance" transfer by ensuring smooth connection with urban public transportation systems such as urban rail transit, buses and taxis, as well as ensuring a layout that results in good transition to airports, long-distance bus stations and passenger ports.

1.0.8 The design of railway passenger station shall meet ecological requirements and comply with regulations on energy conservation, land conservation, water conservation, material conservation and environmental protection.

1.0.9 Reliable new technology, new process, new materials and new equipment shall be adopted for the design of railway passenger station.

1.0.10 In addition to this Code, the design of railway passenger station shall also comply with other relevant current standards of China.

2 Terms

2.0.1 Railway passenger station

The place rendering boarding-alighting service and other services for railway passengers, mainly consisting of station building, facilities for passenger transport, and supporting urban facilities (station square and supporting urban traffic facilities).

2.0.2 Railway passenger station building

Public building rendering service for railway passengers, mainly consisting of entry concourse and exit concourse, waiting area (hall/room), ticket room, passenger traffic service rooms and auxiliary rooms, baggage room, commercial service rooms, etc.

2.0.3 Facilities for passenger transport

A general designation for the buildings, structures (platform, platform canopy, underpass, overpass, etc.) and facilities (ticket gate, lift, escalator, passenger guidance system, etc.) within the range of railway passenger station rendering passenger services.

2.0.4 Terrace for station building

A flat area extending outward from the exterior walls of the railway passenger station building and connecting each part of the station building as well as the entrances and exits.

2.0.5 Concourse

The hall inside the railway passenger station building where the arriving and departing passengers pass by.

2.0.6 Maximum number of passengers gathered in waiting area (hall/room)

Average value of the maximum numbers of passengers (including seeing-off people) in the waiting area (hall/room) of railway passenger station within 8 min~10 min every day (24 hours) in the month with the maximum number of passengers dispatched.

2.0.7 Number of passengers dispatched during peak hour

The number of passengers dispatched at a railway passenger station during peak hour of the day averaged for all the days of the month with the maximum number of passengers dispatched.

2.0.8 Design number of baggage pieces

The daily average number of baggage pieces in the month with the maximum number of baggage pieces handled.

2.0.9 Passenger guidance system

The signage system for guiding passengers in public places.

2.0.10 Green railway passenger station

The railway passenger station in harmonious coexistence with nature, that saves resources (land, energy, water and materials), protects the environment and reduces pollution to the maximum extent within the whole life cycle of the building, providing space for healthy, convenient and efficient services for passengers and staff.

2.0.11 Underground railway passenger station

The railway passenger station with all or part of the station buildings or facilities for passenger transport being located below the ground surface.

2.0.12 Accessible facilities

The safety facilities for people with reduced mobility such as the disabled and the elderly, as well as those with visual impairment.

3 Overall Design

3.1 General Requirements

3.1.1 The location of railway passenger station shall comply with the following requirements:

1 The location of railway passenger station shall be determined in compliance with the requirements of railway transportation and in harmony with the land use planning, urban overall planning and integrated transportation system planning in the region in question.

2 The location of railway passenger station shall be determined through comparison based on such factors as the topographical and geological conditions, distribution of existing buildings, land resources development and urban development.

3 The location of railway passenger station shall achieve coordination with other means of transport and facilitate the mobility of passengers.

3.1.2 The design of railway passenger station shall ensure the intensive utilization of land resources and proper commercial development.

3.1.3 The station building, facilities for passenger transport and supporting urban facilities shall be subject to unified planning and systematic design.

3.1.4 When the station building is located near the railway line, the requirements of railway construction gauge shall be complied with.

3.1.5 Protective and anti-collision measures should be taken for the columns of buildings or structures resting between tracks.

3.1.6 Green building design shall be conducted for railway passenger stations. The evaluation rating of medium and small stations shall not be lower than "one star" specified in the *Evaluation Standard for Green Railway Passenger Stations* (TB/T 10429), that of large stations shall not be lower than "two star", and that of very large stations shall not be lower than "three star".

3.1.7 The fire protection design for railway passenger stations shall comply with the current standards such as *Code for Fire Protection Design of Buildings* (GB 50016), *Code for Design of Fire Prevention for Railway Engineering* (TB 10063) and *Code for Design of Emergency Evacuation and Rescue Works for Railway Tunnel* (TB 10020).

3.1.8 Repair and maintenance facilities that are safe, reliable, and interconnected shall be provided for station buildings and facilities for passenger transport.

3.1.9 When the station building is integrated with other buildings, the electromechanical systems of different functional zones should be deployed separately.

3.1.10 The layout of underground railway passenger station shall be coordinated with the surface buildings, urban roads, underground pipelines, underground structures, etc. on the premise of meeting the functional requirements, and the design of ventilation, lighting, hygiene, disaster prevention, etc. shall comply with the relevant standards of China.

3.1.11 Modern information technology shall be actively adopted for railway passenger station to raise the level of intelligence and passenger services.

3.1.12 The building information modeling (BIM) technology shall be actively adopted for railway passenger station to realize informatization, intelligence and sustainability during the whole life cycle of the station.

3.2 Station Scale

3.2.1 The scale of passenger railway station shall be determined as per Table 3.2.1-1 and Table 3.2.1-2 according to the maximum number of passengers gathered in waiting area (hall/room) or the number of passengers dispatched during peak hour.

Table 3.2.1-1 Scale of Mixed Traffic Railway Station

Station scale	Maximum number of passengers gathered in waiting area (hall/room) (*H*)
Very large	$H \geq 10\ 000$
Large	$3\ 000 \leq H < 10\ 000$
Medium	$600 < H < 3\ 000$
Small	$H \leq 600$

Table 3.2.1-2 Scale of High-speed Railway and Intercity Railway Passenger Stations

Station scale	Number of passengers dispatched during peak hour (pH)
Very large	$\mathrm{pH} \geq 10\ 000$
Large	$5\ 000 \leq \mathrm{pH} < 10\ 000$
Medium	$1\ 000 \leq \mathrm{pH} < 5\ 000$
Small	$\mathrm{pH} < 1\ 000$

3.2.2 The construction area of station building shall be determined through calculation based on the maximum number of passengers gathered in waiting area (hall/room) and as per the following indexes:

1 The construction area of buildings of medium and small railway passenger stations should be 5 m^2/person～8 m^2/person.

2 The construction area of buildings of very large and large railway passenger stations should be 8 m^2/person～15 m^2/person.

3 For the station buildings with transfer space and urban corridors connected with other modes of transportation, the construction area of the transfer space and urban corridors may be determined according to the actual situation.

3.2.3 The area of supporting traffic facilities such as urban rail transit station, bus station, and parking lots for private vehicles and taxis, as well as the commercial facilities outside the railway passenger station shall not be included in the construction area of the railway passenger station building.

3.2.4 When the construction area of railway passenger station building is calculated according to the *Calculation Code for Construction Area of Building* (GB/T 50353), the following items shall not be included in the construction area:

1 The open space below the elevated waiting hall.

2 The separate overpass or underpass for passengers crossing the station and yard.

3 The passenger passage between the main platform and the line-side station building below the track level.

4 Train operation room and cleaning room below the overpass, or its staircases/escalators leading to the platform.

3.3 Interface Design

3.3.1 The railway passenger station shall be designed in coordination with supporting urban traffic facilities in terms of functional configuration, passenger flow line, elevation, space, etc., and shall also be coordinated with the municipal heat supply, gas supply, electricity supply, water supply and drainage, information system, etc.

3.3.2 The railway passenger station shall be designed in coordination with related works such as station and yard, earthworks, bridge, tunnel and OCS.

3.3.3 All the equipment pipelines at the railway passenger station shall be planned as a whole. Underground pipe culverts should be constructed when various outdoor underground pipelines are present; the indoor equipment pipelines shall be coordinated to facilitate laying and maintenance.

4 General Plan

4.1 General Layout

4.1.1 General layout of railway passenger station shall comply with the following requirements:

1 The flow line and functional layout of railway passenger station shall be designed to facilitate boarding-alighting and untwining of passengers.

2 The railway passenger station shall be smoothly connected with urban rail transit and roads.

3 The functional layout, elevation design, traffic organization and landscape design shall be coordinated with the urban planning.

4 The general layout shall be conducive to the intensive use of land resources, with a certain margin reserved for future development.

4.1.2 The design elevation of the site for railway passenger station shall comply with the national and local standards on flood control and waterlogging prevention, and shall be coordinated with the local regulatory planning.

4.1.3 The form of station building shall be determined according to the track conditions, topographic conditions, station area planning, urban supporting facilities, comprehensive development, mode of operation management, etc.

4.1.4 The flow line design of railway passenger station shall comply with the following requirements:

1 The entry, exit and transfer flow lines for passengers shall be short and convenient.

2 For very large and large passenger stations, the entry flow lines and exit flow lines for passengers shall be separated.

3 The passenger flow lines should be separated from flow lines of vehicles, baggage and mails, avoiding intersection.

4.1.5 The terrace for station building shall be provided, which shall comply with the following requirements:

1 The length of the terrace shall not be less than the total length of the main station building.

2 The width of the terrace connected with the station square should not be less than 35 m for very large stations, 25 m for large stations and 10 m for medium and small stations.

3 For railway passenger station with multi-floor traffic mode, the terrace for station building shall be designed for different floors and the width of the terrace for each floor should not be less than 10 m.

4.1.6 The garbage collection facilities, garbage transfer centers and cross-line facilities shall be arranged as needed at railway passenger station.

4.1.7 Dedicated parking lot shall be arranged near the baggage office and a baggage transport corridor shall be provided between the parking lot and the office.

4.1.8 A parking lot for railway staff shall be arranged for the railway passenger station and its scale may be determined through calculation based on the service demand and the local standards for parking lot design.

4.2 Urban Supporting Facilities

4.2.1 The design of station square shall comply with the following requirements:

1 The pedestrian flow line and the vehicle flow line shall be separated and shall be conducive to the traffic organization within the passenger station and the connection with external roads.

2 Hardened antiskid ground surface shall be adopted and shall comply with the drainage requirements.

3 For passenger stations with a large passenger volume in specific seasons or holidays, the station square shall be able to accommodate temporary waiting facilities.

4 Where the roads on the station square are located close to the terrace for station building or other areas where people are crowded, anti-collision facilities shall be provided.

5 The buildings and structures on the station square shall not affect the architectural aspects of the station building.

4.2.2 The design of pedestrian zone in the station square shall comply with the following requirements:

1 The pedestrian zone shall be connected to the bus stations, urban rail transit stations, taxi zones and other traffic facilities.

2 The ground surface shall be 0.15 m above the motor way surface.

3 The area of the pedestrian zone should be determined through calculation as per 1.83 m^2/person based on the maximum number of passengers gathered in waiting area (hall/room).

4 Seats and other related service facilities shall be provided.

4.2.3 Toilets shall be designed on station square and the minimum usable floor area may be determined as per the maximum number of passengers gathered in waiting area (hall/room). The minimum usable floor area shall be determined as per 25 m^2 or four toilet cubicles for every thousand persons. For large station square, toilets should be distributed properly.

4.2.4 The station square shall be provided with management room according to the relevant national or local regulations.

4.2.5 The greening rate of station square should not be less than 10% and the greening and landscape design shall comply with functional and environmental requirements.

4.2.6 Entrances and exits facing different directions should be designed and multi-floor transportation should be adopted for the very large and large railway passenger stations.

4.2.7 The transfer distance between railway passenger station and urban transport stops should not be greater than 300 m.

4.2.8 The scale of the site for urban traffic supporting facilities such as bus stations, taxi stands and parking lots for private cars shall be determined according to the traffic volume, with proper allowance being made. The taxi pick-up area and drop-off area shall be arranged separately according to the passenger flow line.

4.2.9 The design of curbside of roadway for cars shall comply with the following requirements:

1 The unit length of roadway curbside for cars should be 7 m.

2 The capacity of roadway curbside for cars shall be determined through calculation based on the number of passengers carried by a car and the average stopping time of cars. The average number of passengers carried by taxis should be determined as per 1.5 persons/taxi and that carried by private vehicles should be determined as per 2.5 persons/vehicle. The drop-off time and pick-up time of cars should be determined as per 20 s ～ 40 s and 6 s ～ 26 s respectively.

3 The length of the roadway curbside in the pick-up area shall also meet the demands of traffic capacity as well as traffic organization and management.

5 Station Building

5.1 General Requirements

5.1.1 The station building shall be divided into public area, office area and equipment area and the following requirements shall be complied with:

1 The public area should be spacious, with orderly passenger flow lines. The evacuation in the public area must comply with the *Code for Fire Protection Design of Buildings* (GB 50016).

2 The office area should be arranged compactly and channels connecting the office area to the public area shall be provided.

3 The equipment area should be set far away from the public area and should efficiently utilize the space of the building.

5.1.2 The design of passenger flow line of railway passenger station shall comply with the following requirements:

1 The entry, exit and transfer flow lines shall be separated for large and very large passenger stations, should be separated for medium passenger stations and may be combined for small passenger stations.

2 The entry flow line may be designed based on the following sequence: ticket purchasing, identity verification, security check, waiting, and check-in.

3 The check-out facilities shall be provided on the exit flow line.

4 The transfer flow line should be designed based on intra-station transfer.

5.1.3 The width of access, transfer passages and staircases at railway passenger station shall not only meet the demand of peak-hour passenger volume, but also comply with the *Code for Fire Protection Design of Buildings* (GB 50016).

5.1.4 The depth of the railway passenger station building beside lines shall be enough for the entry of passengers into the station.

5.1.5 The railway passenger station rendering baggage and express delivery services shall be provided with storehouses, sites and supporting facilities, which shall facilitate the handling of baggage and express delivery business. The flow line of baggage and express delivery goods shall be separated from the passenger flow line.

5.1.6 The maintenance facilities for station building shall meet the requirements for inspection and maintenance of the building and shall ensure work safety. When mobile inspection and maintenance equipment is adopted, the passing conditions, working space and ground load capacity shall be allowed for.

5.1.7 In the design of station building, the BIM technology shall be encouraged to optimize the building functions and achieve interdisciplinary collaboration, so as to enhance coordination and system optimization.

5.2 Concourse

5.2.1 Entry concourse and exit concourse shall be designed in the station building. For small stations, the entry concourse should be integrated with the waiting area (hall/room), its usable floor area shall not be less than 250 m^2; the usable floor area of exit concourse should not be less than 150 m^2; when the entry concourse is integrated with the exit concourse, the usable floor area shall not be less than 350 m^2. For medium or larger stations, the usable floor area of entry concourse and exit concourse shall be determined based on the number of passengers dispatched during peak hour; the usable floor area of entry concourse shall not be less than 0.25 m^2/person and that of exit concourse should not be less than 0.2 m^2/person.

5.2.2 Service facilities such as information center and left-luggage office shall be designed inside the entry concourse. The self-service package lockers should be provided for medium or larger passenger stations. Toilets and excess fare room shall be provided inside the exit concourse.

5.2.3 Security check zone shall be arranged at the main entrances of arrival concourse in the station building. The usable floor area of the security check zone shall be determined according to the quantity and layout of security check equipment as well as the security check requirements. The minimum usable floor area of each security check zone shall accommodate two sets of security check equipment.

5.2.4 Identity verification facilities shall be provided in the entry concourse.

5.3 Waiting Area (Hall/Room)

5.3.1 The total usable floor area of waiting area (hall/room) shall be determined as per the maximum number of passengers gathered in waiting area (hall/room) and shall not be less than 1.2 m^2/person. The usable floor area of waiting area (hall/room) of very large and large railway passenger stations shall be increased by 5% based on the calculated value.

5.3.2 Soft-seat waiting area should be arranged at very large and large railway passenger stations according to the requirements of passenger transport. The number of waiting passengers in the soft-seat waiting area may be 4% of the maximum number of passengers gathered in waiting area (hall/room) for mixed traffic railway, or may be 10% of that for high-speed railway and intercity railway. The usable floor area of the soft-seat waiting area shall not be less than 2 m^2/person.

5.3.3 Accessible waiting area shall be arranged at medium or larger passenger stations and sites dedicated to wheelchairs shall be provided in waiting area of small passenger stations. The design of accessible waiting area shall comply with the following requirements:

1 The accessible waiting area should be close to the entry gate and the accessible lift.

2 The number of waiting passengers in the accessible waiting area may be 4% of the maximum number of passengers gathered in waiting area (hall/room) and the accessible waiting area shall not be less than 2 m^2/person.

5.3.4 Business-class waiting rooms may be arranged at railway passenger station as required. The design of business-class waiting rooms should comply with the following requirements:

1 Separate gate and a road directly connected to station square shall be provided.

2 Separate identity verification and security check facilities shall be provided.

3 Toilets, washroom, attendants' room and spare parts room shall be provided. Hot water shall be supplied in the washroom.

5.3.5 Seats in the ordinary waiting area (hall/room) shall be arranged to facilitate passengers going to the entry gate. The clear width of the passage between seats shall not be less than 1.3 m and shall meet the requirements for the waiting area dedicated to military men (group).

5.4 Ticketing Room

5.4.1 The ticketing rooms shall include the ticket hall, ticket office, bill room, management office, etc. The income office shall be designed for medium or larger passenger stations and the ledger office shall be designed for very large and large passenger stations.

5.4.2 The ticketing rooms shall comply with the following requirements:

1 The ticket room for very large and large passenger stations shall be located near the entrance of the station building and that for medium and small passenger stations may be located near the waiting area of the station building.

2 Ticket vending (fetching) machine may be provided at railway passenger stations as required and independent ticket vending (fetching) hall may be designed when necessary.

3 The ID card certification counter (s) shall be designed in the ticket hall.

4 Toilets for ticket sales staff shall be designed in the ticket office area.

5.4.3 The number of ticket counters shall be determined according to the number of booking office machines and 1 ~ 2 ticket counters should be reserved in each ticket hall.

5.4.4 The ticket counter shall comply with the following requirements:

1 The distance between centers of adjacent ticket counters should be 1.6 m and the distance between the center of ticket counter and the wall edge should not be less than 1.2 m.

2 The height from the floor of the ticket hall to the sill of ticket counter shall be 1.0 m.

3 The design of accessible ticket counter (s) shall comply with Article 12.0.6 of this Code.

5.4.5 Automatic ticket vending (fetching) machines shall be arranged compactly in the ticket hall and should be installed in the wall. Enough space shall be provided for queuing-up and for maintenance.

5.4.6 The depth of the ticket hall should not be less than 13 m for very large stations, 11 m for large stations and 9 m for medium and small stations.

5.4.7 The ticket office shall comply with the following requirements:

1 The area for each ticket counter shall not be less than 6 m^2.

2 The area of the ticket office shall not be less than 14 m^2.

3 No door shall be designed between the ticket office and the public area.

4 The floor of the areas in the ticket office other than the working area of accessible ticket counter(s) should be 0.2 m higher than that of the ticket hall and the antistatic raised floor shall be adopted.

5 Anti-theft facilities shall be provided in the ticket office.

5.4.8 The bill room shall comply with the following requirements:

1 The area of bill room should not be less than 15 m^2 for medium and small passenger stations, and for very large and large passenger stations, a bill room with area not less than 30 m^2 shall be designed for each ticket room.

2 The bill room shall be provided with moisture-proof, rat-proof, anti-theft and alarm facilities.

5.5 Other Service Facilities

5.5.1 The information and reception counter may be provided in the concourse or waiting area as required.

5.5.2 Commercial facilities shall be designed in the station building according to the scale of the station and the entry and exit flow lines. The scale of in-station commercial facilities should be 2% ~ 4% of the construction area of the station building for small passenger stations, 4% ~ 8% for medium passenger stations and 8% ~ 10% for very large and large passenger stations. The scale of commercial facilities for stations with commercial potentials may be determined as needed. The setting of commercial facilities inside the station shall comply with the *Code for Design on Fire Prevention for Railway Engineering* (TB 10063).

5.5.3 In addition to *Standard for Design of Urban Public Toilets* (CJJ 14), the setting of toilets for passengers shall also comply with the following requirements:

1 The position and signs of the toilets shall be easily identified by passengers.

2 The number of toilet cubicles should be determined as per 2.5 cubicles/100 persons based on the maximum number of passengers gathered in waiting area (hall/room). The proportion of toilet cubicles for males and females shall be 1 : 2. The number of toilet bowls in male toilets shall not be less than three and that for female toilets shall not be less than four. At least one pedestal pan shall be provided in each toilet. In addition, the toilet for males shall be provided with the same number of urinals as toilet bowls.

3 Boards and hooks shall be provided in the toilet cubicles.

4 Washrooms should be designed in both male and female toilets and mirrors shall be provided in washrooms. The number of faucets shall be determined as per 1 faucet/150 persons based on the maximum number of passengers gathered in waiting area (hall/room) and shall not be less than three.

5 The distance from any point in the waiting area (hall/room) to the nearest toilet should not be longer than 50 m and toilets shall be distributed properly for large and very large stations.

6 The length and width of a toilet cubicle shall not be less than 1.5 m and 1.0 m respectively. The clear distance between two rows of toilet cubicles along two opposite sides of the toilet room shall not be less than 2.0 m and the clear distance from toilet cubicles arranged along only one side of the toilet room to the opposite wall or urinals shall not be less than 2.0 m.

7 The toilet layout shall meet the privacy requirement.

8 For the station building provided with transfer area, toilets and washrooms shall be arranged in the transfer area according to the volume of transfer passengers.

9 The setting of accessible toilet shall comply with Article 12.0.11 of this Code.

10 Independent cleaning room shall be provided in the toilets.

11 The unisex toilet shall be provided according to *Standard for Design of Urban Public Toilets* (CJJ 14).

5.5.4 Water heater room for passengers shall be set separately in the station building and shall be isolated from toilets.

5.5.5 The service facilities for mothers and infants shall be arranged in the railway passenger station building and shall comply with the following requirements:

1 Independent infant-caring room shall be arranged and waiting area dedicated to mothers and infants should be arranged at very large, large and medium passenger stations; independent infant-caring room should be arranged at small stations.

2 The usable floor area of the infant-caring room shall not be less than 10 m^2.

3 Facilities for protecting breastfeeding privacy shall be provided in the infant-caring room and the ground of the room shall be skid proof.

4 The infant-caring room shall be equipped with infant care table, wash basin, crib, seats, etc., and should be provided with constant-temperature air conditioner and calling device.

5.6 Passenger Transport Rooms and Auxiliary Rooms

5.6.1 Passenger transport rooms such as passenger transport duty office, excess fare room, police duty office, control (broadcasting) room, water supply staff room, sewage discharge staff room, water heater room, security check office and other areas related to operation should be arranged according to the service requirements, and shall comply with the following requirements:

1 The excess fare room shall be located near the waiting area (hall/room), exit or ticket gate, and its usable floor area shall not be less than 2 m^2/person based on the number of persons in the maximum shift, and should not be less than 10 m^2. Anti-theft facilities shall be provided in excess fare rooms depositing money and bills.

2 The water supply staff room and the sewage discharge staff room shall be arranged separately, and their usable floor area shall not be less than 3 m^2/person based on the number of persons in the maximum shift, and should not be less than 8 m^2.

3 Police duty office shall be provided inside the passenger station where passengers are concentrated and its usable floor area should not be less than 25 m^2.

5.6.2 Work shift room, break room, changing room, staff club, bathroom, dining room, cleaning room (including tool room), etc. shall be arranged at railway passenger station as required and shall comply with the following requirements:

1 The work shift room shall be arranged for medium or larger passenger stations and its usable floor area shall not be less than 1 m^2/person based on the number of persons in the maximum shift, and should not be less than 30 m^2.

2 The usable floor area of break room shall not be less than 4 m^2/person based on 2/3 of the number of persons in the maximum shift, and should not be less than 20 m^2.

3 The usable floor area of changing room shall not be less than 1 m^2/person based on the number of persons in the maximum shift.

4 Staff club, bathroom and dining room should be arranged at medium or larger passenger stations.

5 The cleaning tool room should be arranged in the public area of medium or larger passenger stations. Equipment storeroom shall be designed when mechanical cleaning is adopted.

6 Toilets and washrooms shall be arranged near the baggage handling place and staff workplace at medium or larger passenger stations.

5.6.3 The passenger traffic management rooms should include stationmaster's office, conference room, office, duty room, spare parts room, etc., and shall comply with *Design Code for Office Building* (JGJ 67).

5.6.4 Technical operation rooms shall be arranged at railway passenger station as required. They shall be arranged compactly by categories and the high-voltage and low-voltage interference shall be avoided.

5.6.5 The control room should be integrated with train operation room at high speed railway and intercity railway passenger stations, and should be close to the fire control room. The control room may be integrated with ticket checking office at small railway passenger stations.

5.6.6 The temporary storeroom for prohibited goods shall be arranged at railway passenger station.

5.7 Baggage Room

5.7.1 Baggage office should be designed at mixed traffic railway stations. Baggage deposit office and reception office should be separated according to the entry and exit flow lines at very large and large passenger stations, and may be combined at medium passenger stations. Baggage post may be arranged at small passenger stations.

5.7.2 The baggage corridor shall be provided at passenger stations providing baggage service. The baggage storehouse at very large and large passenger stations should be connected to the cross-line baggage tunnel.

5.7.3 The main composition of baggage rooms shall comply with Table 5.7.3.

Table 5.7.3 Composition of Baggage Rooms

Room	Design number of baggage pieces N (pcs)			
	$N \geqslant 2\,000$	$1\,000 \leqslant N < 2\,000$	$400 \leqslant N < 1\,000$	$N < 400$
Baggage storehouse	Shall be designed	Shall be designed	Shall be designed	Shall be designed
Baggage deposit hall and reception hall	Shall be designed	Shall be designed	Shall be designed	Shall be designed
Office	Shall be designed	Shall be designed	Shall be designed	Should be designed
Bill room	Shall be designed	Shall be designed	Should be designed	Not designed
General inspection room	Shall be designed	Not designed	Not designed	Not designed
Stevedore rest room	Shall be designed	Shall be designed	Should be designed	Not designed
Traction car shed	Shall be designed	Shall be designed	Should be designed	Should be designed
Trailer store room	Shall be designed	Should be designed	Should be designed	Not designed

Note: N refers to the daily average number of all the baggage pieces (including those sent, transferred and received) in the month with the maximum number of baggage pieces in the recent statistical year, considering turnover and future development.

5.7.4 The baggage storehouse shall comply with the following requirements:

1 The departure, arrival and transit baggage storehouses at very large and large passenger stations should be arranged separately.

2 Lifting facilities shall be provided for the baggage storehouse below track level and the multi-storey baggage storehouse, and shall be able to accommodate one baggage trailer.

3 Ramp for baggage trailers connecting different stories of baggage storehouse at very large passenger stations shall be provided. When there is a height difference between different baggage operation areas at railway passenger station and between the baggage operation area and the platform or station square, a ramp for small handling equipment shall be reserved. The gradient of the ramp shall not be greater than 1 : 12. The clear width of the ramp shall not be less than 3 m when handrails are installed and shall not be less than 4 m when no handrail is installed.

4 Conveyor belt may be provided in the luggage reception hall at very large passenger stations.

5.7.5 The usable floor area of baggage storehouse shall be calculated according to the number of baggage pieces.

5.7.6 Outdoor storage yard should be reserved for the railway passenger stations with a design number of baggage pieces exceeding 2 000 pcs and rainproof facilities shall be provided in the yard.

5.7.7 Unclaimed baggage storeroom should be designed at very large or large passenger stations and its usable floor area may be determined according to 1% of the design number of baggage pieces and should not be less than 20 m^2.

5.7.8 The clear height inside the baggage storehouse shall not be less than 4 m. The baggage storehouse with mechanical operation shall meet the requirements for mechanical operation and the width and height of the door shall not be less than 3 m.

5.7.9 The baggage storehouse should be provided with high windows and protection facilities.

5.7.10 The usable floor area of baggage deposit hall and reception hall and the number of counters for deposit and reception shall not be less than those specified in Table 5.7.10.

Table 5.7.10 Usable Floor Area of Baggage Deposit Hall and Reception Hall and Number of Counters for Deposit and Reception

Item	Design number of baggage pieces N (pcs)					
	$N<600$	$600\leqslant N<1\,000$	$1\,000\leqslant N<2\,000$	$2\,000\leqslant N<4\,000$	$4\,000\leqslant N<10\,000$	$N\geqslant 10\,000$
Number of baggage counters	1	1	2	4	7	10
Usable floor area of baggage deposit hall and reception hall (m^2)	15	25	30	60	150	300

5.7.11 A passage at least 1.5 m wide shall be provided between the baggage storehouse and the baggage deposit hall and reception hall and the passage shall be provided with gate that can be opened.

5.7.12 Baggage scanners shall be provided at medium or larger passenger stations as required.

5.7.13 Anti-collision measures shall be taken for walls and columns in the baggage operation area and on both sides of the baggage transport corridor.

5.8 Indoor Environment

5.8.1 The space inside the station building shall be sufficiently unobstructed, open and bright and the dimensions of the space shall meet different functional requirements.

5.8.2 The main spatial design of railway passenger station shall be able to provide guidance for the passengers and shall facilitate the identification of different functional units and the evacuation of passengers.

5.8.3 The requirements on the indoor environment of station building such as daylighting, ventilation, heat preservation and insulation, sound insulation, pollutant control, etc. shall comply with relevant current standards.

5.8.4 Natural lighting and ventilation should be utilized for the public area of station building and measures shall be taken to reduce dazzle light in daylighting design.

5.8.5 The indoor acoustic design of station building shall comply with the following requirements:

1 Acoustic design should be adopted for the station building with a public area of 50 000 m^2 or with an average height of more than 18 m.

2 Measures for noise and vibration reduction shall be taken for the main noise and vibration sources and measures for sound insulation, sound absorption and vibration isolation shall be taken for the equipment rooms that may produce noise and vibration.

3 The reverberation time at the frequency of 500 Hz in the public area of station building should comply with Table 5.8.5.

Table 5.8.5 Reverberation Time at Frequency of 500 Hz in Public Area with Different Capacities

Capacity ($1\times10\,000\ m^3$)	$\leqslant 100$	>100
Reverberation time (s)	$\leqslant 4$	$\leqslant 5.5$

5.9 Interior Decoration and Details

5.9.1 The interior decoration of station building shall comply with the following requirements:

1 It shall be in harmony with the architectural form.

2 It shall ensure safety, durability and economic efficiency, and facilitate repair and maintenance.

3 Fireproof, anticorrosive, environmentally friendly and highly cleanable materials shall be adopted.

4 Facilities such as guidance signs, advertising boards and fire-fighting control equipment, as well as the terminal equipment such as water supply & drainage equipment, HVAC, high-voltage and low-voltage installations, shall be designed and planned in coordination with the interior decoration. The advertising boards shall not interfere with the layout of various signs in the station building.

5 The interior decoration shall comply with the *Code for Fire Prevention in Design of Interior Decoration of Buildings* (GB 50222).

5.9.2 The exposed corners on walls and columns in the indoor public areas should be rounded. The area on the wall surface within the height of 1.80 m should be finished with impact-resisting materials and anti-collision railings shall be provided at 0.10 m above the ground for the glass curtain walls.

5.9.3 The height of breast boards at balconies shall not be less than 1.30 m and the height of handrails shall be 1.10 m. When breast boards made of glass are employed, toughened glass with lamination shall be adopted. Anti-collision structure shall be provided at a height of 0.10 m above the ground.

5.9.4 The toughened glass with lamination shall be adopted for glass partition. Anti-collision facilities shall be provided at the bottom. Anti-collision structure shall be provided at a height of 0.10 m above the ground.

5.9.5 The railings and breast boards of staircases and escalators shall be safe and reliable, with no edges and corners at the ends.

5.9.6 Safe and reliable connection must be provided between the ceiling and the main structure.

5.9.7 The ground of the passenger station building shall be wear-resistant, antiskid, corrosion-resistant and easily cleaned. The slip resistance of the ground shall comply with the *Technical Specification for Slip Resistance of Building Floor* (JGJ/T 331).

5.9.8 Maintenance access and manhole shall be provided on the ceiling of very large and large railway passenger station buildings. Removable ceiling should be designed for medium and small station buildings.

5.10 Curtain Wall and Metal Roofing

5.10.1 The design of curtain wall shall follow the principle of safety, practicability and elegance, and shall facilitate the production, installation and maintenance.

5.10.2 Safety glass shall be adopted for the glass curtain wall. Clear warning signs and anti-collision facilities shall be provided at the parts susceptible to collision.

5.10.3 Stone and glass curtain walls should not be used for the exterior wall above the railway line. If unavoidable, protective facilities such as roof overhangs cornice and shock proof shed shall be provided below the curtain wall.

5.10.4 The thoroughly hidden frame glass curtain wall is prohibited and inverted (faced) stone panel or ceramic tile shall not be adopted above the main passage for passengers and above the railway line.

5.10.5 For the design of metal roofs, the type of metal plates and structure system shall be determined according to the wind and snow loads, structure form, thermal performance, roof slope, etc.

5.10.6 Anti-fall metal nets or other anti-fall devices should be provided for the glass skylight of the metal roof, and shall comply with the *Technical Specification for Skylight and Metal Roof* (JGJ 255).

5.10.7 The design rainwater recurrence interval for metal roofs shall be 100 years for large or very large passenger stations and 50 years for medium and small passenger stations. The metal roofs shall be provided with an overflow system.

5.10.8 Maintenance facilities directly connected to the metal roof shall be provided and the anti-fall devices shall be installed on the metal roof without parapet or with parapet (including the upturning roof cornices) lower than 500 mm.

5.10.9 In severe cold regions, the ice and snow melting devices should be provided for cornices of metal roofs and skylights, as well as water collection and drainage gutters, and snow boards or snow guards should be provided for pitched roofs and arched roofs. For the metal roofs in severe cold and cold regions, measures shall be taken to prevent icicles from forming at roof cornices after melting of snow.

5.11 Building Energy Conservation

5.11.1 The energy efficiency design of station building shall comply with the *Design Standard for Energy Efficiency of Public Buildings* (GB 50189) and *Code for Design of Energy-saving of Railway Engineering* (TB 10016).

5.11.2 In the station building, the location of the energy equipment room shall be determined according to the thermal environment zoning, and the distance for conveying cold water, hot water and air shall be short.

5.11.3 Natural lighting shall be fully utilized for the main functional areas of station building. The area ratio of the daylight factor should not be less than 60% and shall be calculated by Formula (5.11.3):

$$\varphi_{NL}=\frac{F}{F_{all}}\times 100\% \qquad (5.11.3)$$

Where

φ_{NL}——qualified area ratio of daylight factor;

F——total area of main functional areas in the station building meeting the requirements of the *Standard for Daylighting Design of Buildings* (GB 50033) (m^2);

F_{all}——total area of main functional areas in the station building (m^2).

5.11.4 Natural ventilation and cooling shall be utilized for the main functional areas of station building, and the mechanical exhaust device may be provided to strengthen natural air compensation. The ratio of the open area for natural ventilation to the ground area should comply with Table 5.11.4.

Table 5.11.4 Ratio of Open Area for Natural Ventilation to Ground Area φ_{NV}

Climatic region	Severe cold and cold region	Hot-summer cold-winter region	Hot-summer warm-winter region, mild region
Single storey station	$\varphi_{NV}\geqslant 3.0\%$	$\varphi_{NV}\geqslant 4.0\%$	$\varphi_{NV}\geqslant 4.0\%$
Multistorey station	$\varphi_{NV}\geqslant 1.5\%$	$\varphi_{NV}\geqslant 2.0\%$	$\varphi_{NV}\geqslant 2.0\%$

5.11.5 Except for the severe cold region and cold region, sun-shading measures shall be taken for the external windows (including the transparent curtain walls) of the station building in the main direction of sunlight. The external sun-shading measures shall comply with the following requirements:

1 The movable external sun-shading device should be provided for the east and west, and the horizontal external sun-shading device should be provided for the south.

2 Consideration shall be given to ventilation and winter sunshine for the external sun-shading device.

5.11.6 Clean energy should be adopted for railway passenger station and the main building materials should be obtained from local sources.

5.11.7 The external thermal insulation system should be employed for exterior wall thermal insulation of station building, and safe, reliable and technologically-proven insulation materials and structures shall be adopted. The design of thermal bridge of the exterior wall shall comply with the *Code for Thermal Design of Civil Building* (GB 50176).

5.11.8 Porches or double-layer doors shall be designed at the main entrances and exits of the station building provided with air conditioning and heating system. The ratio of the number of porches and double-layer doors to that of outside doors should comply with Table 5.11.8.

Table 5.11.8 Ratio of Number of Porches and Double-layer Doors to That of outside Doors

Region	Ratio of number of porches and double-layer doors to that of outside doors
Severe cold and cold region	$\varphi_w \geq 50\%$, or $\varphi_n + \varphi_s = 100\%$
Hot-summer cold-winter region, hot-summer warm-winter region	$\varphi_w + \varphi_n + \varphi_s \geq 50\%$

Note: φ_w, φ_n and φ_s refer to the ratios of the number of outside porches, the number of inside porches and the number of double-layer doors to that of outside doors, respectively.

5.11.9 The air tightness of external doors and windows of station building shall comply with the current national standard *Graduations and Test Methods of Air Permeability, Water Tightness, Wind Load Resistance Performance for Building External Windows and Doors* (GB/T 7106).

5.11.10 When the full-glass curtain walls are adopted at the main entrances and exits of station building, the area of the non-hollow glass in the full-glass curtain wall shall not be greater than 15% of light transmission area (doors, windows and glass curtain walls) of the same side of the building. In addition, the mean heat transfer coefficient obtained from weighted calculation shall comply with the current national standard *Design Standard for Energy Efficiency of Public Buildings* (GB 50189).

5.12 Port Station

5.12.1 Passenger transport and port inspection facilities shall be provided at port stations.

5.12.2 The station building and the facilities for passenger transport shall comply with the following requirements:

1 The station building and the facilities for passenger transport for border-crossing service and those for domestic service shall be separated. If border-crossing service and domestic service share the same passage, facilities for separating or isolating border-crossing passengers from domestic passengers shall be provided.

2 Waiting areas for border-crossing passengers may be designed in multi-room form.

3 The setting of waiting room and baggage consignment points for border-crossing passengers shall comply with the regulatory requirements of the inspection authority.

5.12.3 The port inspection facilities shall include frontier inspection facilities, customs inspection facilities, as well as inspection & quarantine facilities. The inspection process may be designed according to the standards for frontier inspection, customs inspection and inspection & quarantine, as well as the actual needs.

5.12.4 Service facilities such as duty-free shop, currency exchange, postal service, tourist information center, reception service and small restaurants may be arranged at port stations.

5.12.5 National flag pole shall be provided at the port station square in the border region.

5.13 Underground Station

5.13.1 The overall layout of underground station shall be determined according to the urban planning, track laying, ambient environment, urban landscape, etc. Measures shall be taken to reduce noise, vibration and impact on ecological environment.

5.13.2 The design of underground station shall meet the requirements of function and traffic volume, in addition, ventilation, lighting, hygiene, waterproofing and disaster prevention measures shall be taken that can ensure boarding-alighting safety and are convenient for management.

5.13.3 The construction area of underground railway passenger station shall cover the main works and auxiliary works, and the construction area of main works and that of auxiliary works should be calculated independently.

5.13.4 The public area, office area and equipment area of underground station should be located on the ground. The office area and the equipment area shall be arranged compactly and the main office area shall be arranged in a centralized manner. The fire pump room should be located near the main passage or fire-fighting passage in the office area or attended equipment area.

5.13.5 The equipment area and the unattended office area located at both ends of the station platform may stretch into the range of calculated length of the platform. However, the stretching length shall not be greater than the length of a car and the distance from the end of the equipment area and the unattended office area to the staircase, escalator or passage shall not be less than 8 m.

5.13.6 The elevation design of underground station shall be determined according to the system functions, construction methods, geological conditions, equipment requirements and other requirements.

5.13.7 The clear width and clear height of the main parts of underground station building shall not be less than those specified in Table 5.13.7-1 and Table 5.13.7-2.

Table 5.13.7-1 Minimum Clear Width of Main Parts of Underground Station

Location	Minimum clear width (m)
One-way stairs in public area	1.8
Two-way stairs in public area	2.4
Stairs arranged in parallel with escalator in public area	1.6
Working stairs (doubling as escape stairs) from platform to track area	1.1

Table 5.13.7-2 Minimum Clear Height of Main Parts of Underground Station

Location	Minimum clear height (m)
Public area in station hall	3.0
Public area on platform	3.0
Management rooms on platform and in station hall	2.5
Station entry/exit passageway	3.0

5.13.8 The location of the entrance and exit of underground station shall facilitate the gathering and evacuation of passengers. The setting of ventilation pavilion shall not only meet the functional requirements, but also meet the requirements of planning, environmental protection and urban landscape. For the design of entrance and exit, ventilation pavilion and cooling tower of underground station, *Code for Design of Metro* (GB 50157) may be referred to.

5.13.9 The design elevation of the ground entrance and exit, emergency exit, ventilation pavilion and other openings shall not only meet the requirements in planning, but also meet the requirements of urban or regional ponding (flood) control water level. If such requirements cannot be met, necessary flood prevention measures shall be taken.

5.13.10 The escalator and lift shall not be installed at the inducing joints and deformation joints of civil structures.

5.13.11 The air ducts above tracks of underground station shall be of reinforced concrete, with structural durability consistent with that of the main structure. Other components above the railway line must be fixed firmly and reliably.

6 Facilities for Passenger Transport

6.1 Platform and Canopy

6.1.1 The length, width and height of station platform shall comply with the *Code for Design of Railway Station and Terminal* (TB 10099).

6.1.2 The distance from the edge of the platform entrance/exit or of the buildings on platform to the line-side edge of the platform shall not be less than 3.0 m, and for medium and small stations under difficult conditions, shall not be less than 2.5 m; for upgraded stations, it shall not be less than 2.0 m.

6.1.3 The platform surface shall comply with the following requirements:

1 Rigid anti-skid floor shall be adopted, which shall be able to sustain the loads of baggage and mail vehicles. Platform accessible to fire trucks shall also be able to sustain the loads of fire trucks.

2 The transverse slope of the platform surface shall not be greater than 1%.

3 The platform at which passenger trains stop shall be provided with 0.10 m-wide warning line along its full length.

6.1.4 The design of platform canopy shall comply with the following requirements:

1 Canopy shall be installed on the station platform and the length of the canopy should be the same as that of the platform. For small stations, the length of the canopy may be determined according to the traffic volume and other demands.

2 The form and height of canopy shall be able to prevent the rain and snow from falling onto the platform. For canopies with columns resting between lines, the width of the opening between two canopies and the height of the cornice shall be able to prevent the rain and snow from falling onto the platform.

3 The spacing between the canopy components and the track shall comply with the current national standard *Structure Gauge for Standard Gauge Railways* (GB 146.2).

4 For the platform accessible to fire trucks, the height from the platform surface to the lowest point of the objects suspended from the canopy shall not be less than 4.00 m.

5 Canopies along the flow line for passenger entry and exit shall be continuous.

6 If no platform canopy covers the entrance or exit of the underpass, a separate canopy shall be provided which should be of enclosed type, with the coverage beyond the entrance or exit by a length not less than 4.00 m.

7 Canopy with columns on the platform should be adopted for medium and small passenger stations and canopy with columns resting between lines may be adopted for very large and large passenger stations. The form of the canopy shall be coordinated with the architectural form of the station building.

6.1.5 The railings (breast boards) on the station platform shall comply with the following requirements:

1 The railings or breast boards near the platform edge shall not be lower than 1.30 m. When the height of platform on the side of the railings or breast boards is greater than 12 m, the railings or breast boards shall not be lower than 2.20 m. There shall be no gap within the height of 0.10 m from the platform surface. When breast boards are not installed directly to the ground, a 0.15 m-wide and

0.10 m-high retaining stand may be provided.

2 The railings at the ends of platform (perpendicular to the line direction) shall not be lower than 1.30 m.

3 When the line-side station building with the same level of track is connected to the platform, the railings or breast boards near the outer edge at the side of the station building shall not be lower than 2.20 m.

6.1.6 The staff room and cleaning room should be arranged centrally on the station platform in combination with the staircase and escalator. Water and power facilities shall also be provided.

6.1.7 Seats may be provided on the station platform.

6.1.8 For very large and large railway passenger stations, the staircase, lift and escalator leading to cross-track facilities shall be arranged on the main platform.

6.2 Cross-track Facilities

6.2.1 The layout of access for passengers shall be determined based on the number of passengers, layout of the station building, entry & exit flow lines, etc., and shall comply with the *Code for Design of Railway Station and Terminal* (TB 10099).

6.2.2 The clear width and clear height of access for passengers shall be determined through calculation and shall comply with Table 6.2.2.

Table 6.2.2 Minimum Clear Width and Clear Height of Access (m)

Item	Very Large Stations	Large Stations	Medium and Small Stations
Minimum clear width	12	8～12	6～8
Minimum clear height of underpass	3.0		2.5
Minimum clear height of enclosed overpass	3.5		3.0

6.2.3 The width of the access of overpass and underpass leading to the platform shall comply with the following requirements:

1 The overpass and underpass should be provided with two-way entrances/exits leading to the platform. The width of access for high-speed railway and mixed traffic railway shall comply with Table 6.2.3-1, and that for inter-city railway shall comply with Table 6.2.3-2. When escalators or lifts are provided at the entrance/exit, the width of access shall be determined according to the number and requirements of the escalators or lifts.

Table 6.2.3-1 Width of Access of Platform for High-speed Railway and Mixed Traffic Railway (m)

Platform	Very Large and Large Stations	Medium Stations	Small Stations
Main platform and island-type intermediate platform	5.0～5.5	4.0～5.0	3.5～4.0
Side intermediate platform	5.0	4.0	3.5～4.0

Table 6.2.3-2 Width of Access of Platform for Inter-city Railway (m)

Name	Medium Stations	Small Stations
Platform	4.5～5.0	3.0～4.0

2 For station upgrading, the existing station access may be used and Item 1 of this article shall be complied with.

6.2.4 Service underpass connected to the platform shall be provided according to the requirements of delivering baggage, mails, foods and beverage, transferring garbage, as well as operation of cleaning and maintenance equipment. The service underpass shall comply with the following requirements:

1 At least one service underpass shall be provided at very large and large passenger stations, and one service underpass may be provided at the medium station undertaking arrival and departure service of passenger trains.

2 The clear width of the underpass shall not be less than 5. 2 m and the clear height shall not be less than 3. 0 m.

3 One entrance/exit should be designed for each platform, which should be located on the end of the platform. The clear width of entrance/exit shall not be less than 4. 5 m, and when conditions do not allow for a width of 4. 5 m and guidance signs are provided at the entrance/exit, the clear width of entrance/exit shall not be less than 3. 5 m.

6. 2. 5 The distance between any two stairs/escalators/ramps of the overpass/underpass on the platform shall comply with the following requirements:

1 The distance should not be less than 20 m for very large and large passenger stations.

2 The distance should not be less than 15 m for medium and small passenger stations.

3 When two escalators are arranged in opposite directions, the distance between their entrances/exits shall meet the requirements on the spacing between operating points of the escalators.

6. 2. 6 Stairs and ramps at the entrance and exit of overpass and underpass shall comply with the following requirements:

1 When only stairs are arranged at the entrance and exit of overpass and underpass for passengers, the steps should not be higher than 0. 14 m and should not be narrower than 0. 32 m; when stairs at the underpass and overpass for passengers are arranged together with the escalators, the stair steps should not be higher than 0. 15 m and should not be narrower than 0. 30 m. There shall not be more than 18 steps in each stair section and the width of platform for straight stairs should not be less than 1. 50 m.

2 Anti-skid measures shall be taken for ramps at the overpass and underpass and the ramp slope should not be greater than 1 : 8.

3 The slope of the ramp at the entrance and exit of the underpass for baggage should not be greater than 1 : 12 and the horizontal distance from the starting point of the ramp to the underpass should not be less than 10 m.

4 The connection between the underpass and the entrance or exit should be of rounded corners.

6. 2. 7 The underpass shall comply with the following requirements:

1 The platform surface at the underpass entrance and exit shall be 0. 02 m above the other parts of the platform surface, with gradual slope as transition.

2 Waterproof and drainage facilities shall be provided for the underpass.

3 The poorly ventilated underpass shall be provided with ventilation facilities and moisture-proof measures shall be taken.

6. 2. 8 The overpass shall comply with the following requirements:

1 Ceiling shall be provided for the overpass. The overpass in severe cold and cold regions shall be enclosed and the retaining structure should be arranged on both sides of the overpass in other regions.

2 The clear height of the railings or breast boards of the overpass shall not be less than 2. 2 m. Railings, breast boards or enclosing structure for the trussed overpass shall be arranged at the inner side of the truss.

3 Toughened glass with lamination shall be adopted when glass windows are arranged on both sides of the overpass for daylighting. Anti-collision measures shall be taken for the French windows made of glass.

4 The overpass shall be designed beautifully on the premise of meeting the requirements of functionality.

6.2.9 Buildings or structures above the railway line shall be of simple form and provided with safe and reliable connection. Moreover, ornamental elements shall not be used and conditions for repair and maintenance shall be reserved.

6.2.10 The opening sashes of windows and of glass curtain walls of elevated waiting hall and passenger overpass shall not be arranged above the main line of high-speed railway.

6.3 Ticket Gate

6.3.1 The number of entry and exit gates shall be calculated based on the passenger volume, through capacity of ticket gate, waiting time, etc.

6.3.2 For the railway passenger station with automatic ticket gates, the manual ticket gate shall be provided beside each group of automatic ticket gates.

6.3.3 The distance from the entry gate to the opposite evacuation door or the stairs leading to the platform should not be less than 4 m, and the distance from the entry gate to the operating point of escalators should not be less than 7 m. The distance from the exit gate to the opposite evacuation door or the stairs should not be less than 5 m, and the distance from the exit gate to the operating point of escalators should not be less than 8 m. For underground stations, the distance from the exit gate to the edge of entrance/exit passageway should not be less than 5 m.

6.3.4 The entry gate and the exit gate shall meet the requirements for safe evacuation and accessibility.

6.3.5 Seats or other facilities that may affect queuing up for ticket checking shall not be arranged near the entry gate and exit gate. In addition, the length of the queuing area in front of the entry gate should not be less than 15 m and that in front of the exit gate should not be less than 7 m.

6.3.6 Two-way check-in and check-out shall be adopted at the ticket gates according to the requirements of transfer.

6.4 Lift and Escalator

6.4.1 The entry and exit passages should be provided with lifts and escalators. The transfer passages with a horizontal transfer distance of more than 300 m should be provided with moving walks, the inclination of which shall not be greater than 12°.

6.4.2 The outdoor escalators should be provided with ceilings and enclosure. Drainage facilities shall be provided for outdoor lifts and escalators.

6.4.3 For escalators, the public service escalators with variable-frequency speed control shall be adopted. Escalators shall comply with the following requirements:

1 For the escalators on the entry and exit passages, the inclination should be 23.2° when stairs are arranged alongside, and may be 23.2° or 27.3° without stairs being arranged alongside.

2 The rated speed of escalators should be 0.5 m/s.

3 The height of one step shall not be less than 0.38 m and the number of horizontal steps shall not be less than three.

6.4.4 The distance from the operating point of escalators to the immobile facilities ahead that affect traffic shall not be less than 8 m; the distance between the operating points of two escalators arranged in opposite directions shall not be less than 16 m. When an escalator is arranged in the direction opposite to the stairs, the distance from the operating point of the escalator to the first step of the stairs shall not be less than 12 m.

6.4.5 Lifts shall comply with the following requirements:

1 The rated capacity of passenger lifts shall not be less than 1 000 kg, and shall not be less than 1 600 kg when the passenger lift is also used for logistics.

2 The rated speed of passenger lifts should be 1 m/s and shall not be less than 0.63 m/s.

3 The gate of the passenger lift should be of two leaves arranged symmetrically, the width of the lift shall not be less than 1 m, and the door of the lift shall not face the railway line.

4 Freight lifts shall comply with the current standards of China.

6.4.6 The height of the escalator handrail shall be neither less than 1.00 m nor greater than 1.10 m. Necessary safety measures shall be taken for escalators with a lifting height of 12 m or above.

6.4.7 The ground at the junction of escalator and platform should be 0.02 m above the platform ground and shall be connected to the platform ground by easy gradient.

6.5 Passenger Guidance System

6.5.1 The passenger guidance system shall be provided at railway passenger station and shall comply with the following requirements:

1 The guidance system shall be arranged at the station square, station building, platform, overpass, underpass, urban corridors, etc.

2 Static or digital signs may be adopted for the guidance system.

3 The guidance system shall be installed safely and continuously at conspicuous positions in a standardized and systematic manner, and shall be easily identified.

4 The guidance systems with different functions shall be distinguished from each other and can be easily identified, and the guidance systems with important functions shall be prioritized when space available is not sufficient.

5 The passenger guidance system of the station shall be coordinated and connected with the municipal guidance system.

6 Coordinated design shall be performed for static and digital signage as well as for the guidance system and the interior and exterior decoration.

6.5.2 Digital signage for entry and exit shall be provided at passenger railway station. The digital signage for entry shall be arranged at the entrance, waiting area (hall/room), ticket hall, entry passage, platform and other areas along the entry flow line; the digital signage for exit shall be arranged at the exit passage, exit concourse and other areas along the exit flow line.

6.5.3 The guidance signs shall be arranged on the main flow lines and shall comply with the following requirements:

1 The entry guidance signs shall start from the bus station, taxi stand, parking lot, urban rail transit station, etc. within the range of the station square, then proceed along the entry flow line all the way to the entrance, waiting area (hall/room), ticket hall, ticket gate, platform, etc.

2 The exit guidance signs shall start from the platform, and then proceed along the exit flow line all the way to the exit passage, exit, etc.

3 The transfer areas and the junctions of roads and passages where multiple directions are available shall be provided with guidance signs. The position and spacing of guidance signs may be determined according to the size of the guidance signs, lighting conditions, spatial arrangement, viewing distance, etc., and shall comply with the *Guidance System for Public Information - Setting Principles and Requirements* (GB/T 15566).

6.5.4 Location signs shall be provided in the following areas of railway passenger station:

1 Station square, bus station, parking lot, taxi stand, urban rail transit station and other supporting facilities within the range of the station.

2 Ticket room, ticket vending (fetching) machine, station entrance, station exit, baggage office, waiting area (hall/room), ticket gate and platform, as well as other facilities for passenger transport.

3 Joint inspection facilities at frontier (port) station for customs inspection, frontier inspection, health quarantine and quarantine of animals and plants, as well as duty-free shop, etc.

4 Passenger service facilities in the public area of the station building, such as information center, police office, left-luggage office, toilets, service facilities for mothers and infants, public phone, drinking water station, automatic teller machine, and commercial areas.

6.5.5 The general layout plan of the station should be provided in the entry concourse, exit concourse and transfer area, and the schematic plan and information index signs for each floor should be provided at the main entrances and exits of the floor in question. The ticket hall, station entrance, waiting area (hall/room) and transfer area of the station building should be provided with information signs.

6.5.6 The setting of station name board on the platform shall comply with the following requirements:

1 At least two suspended station name boards perpendicular to the line direction and two attached station name boards parallel to the line direction shall be installed on each side of the platform with canopies.

2 At least two column-type station name boards or frame-type station name boards parallel to the line direction shall be installed on each side of the platform without canopy.

3 The station name board shall be conspicuous, stable and firmly fixed to the base.

6.5.7 The platforms for EMU (except those only for the departure EMU) shall be provided with ground signs indicating the position of EMU cars.

6.5.8 Station name sign shall face the station square and shall be in harmony with the station building, and its position, font and size shall meet the demand of identification by passengers.

6.5.9 Signs shall be provided for the accessible facilities at passenger railway station and relevant guidance signs shall be provided on the entry and exit flow lines, for which the current national standards *Codes for Accessibility Design* (GB 50763) and *Guidance System for Public Information - Setting Principles and Requirements* (GB/T 15566) shall be complied with.

6.5.10 Safety signs shall be provided at the safety railings of the platform and other conspicuous places of the platform, at the staircases and escalators in crowded places, at positions with elevation changes, as well as at other places with potential dangers, and the current national standard *Safety Signs and Guideline for the Use* (GB 2894) shall be complied with.

6.5.11 "No Smoking" signs shall be provided inside the railway passenger station.

6.5.12 The symbols and other relevant elements on static signs shall comply with the following requirements:

1 Static signs may be designed in the form of graphics, symbols, texts or their combination.

2 The graphics and symbols of static signs shall comply with the current national standard

Public Information Graphical Symbols (GB/T 10001) and other relevant standards.

3 The composition and combination of texts and graphics on the static signs shall comply with the *Public Information Guidance Systems - Design Principles and Requirements for Guidance Elements* (GB/T 20501).

6.5.13 The passenger guidance facilities shall comply with the following requirements:

1 The guidance facilities may be of attached type, suspended type, standing type, column type, table type, frame type, floor type, etc.

2 The base of the guidance facilities shall be able to sustain the loads of the facilities and the connection between the guidance facilities and the base shall meet the relevant safety requirements.

3 Materials harmful or dangerous to human body shall not be adopted for the guidance facilities. Materials of the lighted guidance facilities shall be fireproof and electrical materials shall be insulated.

4 The installation height of static signs shall comply with the following requirements:

1) For the indoor suspended or overhanging sign, the height from the ground to the lower edge of the sign shall not be less than 2.2 m, and for the outdoor suspended or overhanging sign, the height shall be determined according to the requirements for people or vehicles passing below.

2) For the attached guidance sign, the height from the ground to its upper edge shall not be less than 2.0 m.

3) For the attached location sign, the height from the ground to its upper edge should be 1.6 m; when identification at a long distance is required, the height from the ground to its lower edge shall not be less than 2.0 m.

4) The installation height of other types of signs shall comply with the *Guidance System for Public Information - Setting Principles and Requirements* (GB/T 15566).

5 The wear resistance, weather resistance, cleanability and slip resistance of materials of the ground guidance signs shall comply with the *Specification and Test Method for Road Traffic Markings* (GB/T 16311).

6 Reflectorized material or self-luminescent material should be adopted for outdoor safety signs and guidance signs.

7 The guidance facilities shall be convenient for maintenance.

7 Structure

7.1 General Requirements

7.1.1 A safe and applicable structural system that is easy for construction, technologically advanced and economically reasonable shall be used in structural design.

7.1.2 In the cases of underground station and bridge-building integrated station, the design service life of track-bearing storey and the structures below shall be 100 years, and the design service life of the structure of other railway passenger stations shall be 50 years.

7.1.3 The seismic design category and safety grade of structures shall comply with the following requirements:

1 In the cases of underground station, bridge-building integrated station, over-track elevated station building, and other kinds of station buildings where the maximum number of passengers gathered in waiting area (hall/room) is 6 000 or above, the seismic design category of structures shall be Category B and the safety grade shall be Grade Ⅰ.

2 The safety grade of over-track facilities such as canopy and overpass and the metal roofing shall be Grade Ⅰ; in the cases of canopy and overpass as well as other structures with their columns located between tracks, the seismic design category shall be Category B; the seismic design category of the canopy with columns located on platform should be Category C.

7.1.4 The durability design of structures shall comply with the following requirements:

1 In the cases of underground station, bridge-building integrated station, over-track elevated station building, and other kinds of station buildings where the maximum number of passengers gathered in waiting area (hall/room) is 6 000 or above, the main structure and those structural members that cannot be replaced during service life shall be designed based on the durability of 100 years, while the secondary structural members that can be replaced without affecting the railway operation during service life shall be designed based on the durability of 50 years; other kinds of railway passenger stations shall be designed based on the durability of 50 years.

2 The durability of temporary structures should be determined according to the service conditions and the structural features.

7.1.5 The recurrence interval of reference wind pressure and reference snow pressure for the metal roofing of the station, canopy, and overpass should be 100 years.

7.1.6 When the structure is subjected to accidental actions such as fire, hurricane, explosion and collision, structural analysis shall be carried out according to the requirements of relevant current standards and the design of resistance to progressive collapse shall be performed when necessary.

7.1.7 For the connection between railway passenger station and urban rail transit, or between railway passenger station and municipal works, coordinated design shall be performed in respect of the structural layout, settlement control, load transfer, etc.

7.1.8 Overall structure health monitoring should be carried out for the structure of very large station building and may be carried out for the complex structure of large station building. Local structure health monitoring may be carried out for the important members of the station structure that are under severe environment and susceptible to corrosion or subject to alternating load for a long time. The content of structural health monitoring shall be determined based on the risk analysis of important parts of the building structure.

7.2 Loads and Actions

7.2.1 The permanent loads and actions shall include the self-weight of structural elements, building envelope, finishing coat and decoration, fixed equipment and long-term reserve material, as well as earth pressure, hydrostatic pressure, water buoyancy, concrete shrinkage, concrete creep, foundation displacement effect, prestress and other loads to be regarded as permanent loads in design.

7.2.2 The variable loads shall include floor live load, roof live load, equipment load, dust load, wind load, snow load, temperature effect (uniform temperature, gradient temperature, frost-heaving force in severe cold and cold regions), train loads, automobile loads, lateral earth pressure caused by live loads, crowd load, flowing water pressure, ice pressure, construction load, etc.

7.2.3 The accidental actions shall include explosion, collision, fire and other actions that occur accidentally.

7.2.4 The train loads shall be determined according to the specific project conditions and the relevant standards for railway design.

7.2.5 The seismic precautionary intensity shall comply with the basic seismic intensity specified in the national and local standards.

7.2.6 The characteristic value of uniform live load of the crowd on the platform shall be 3.5 kN/m^2, while the factor for combination value, factor for frequent value and factor for quasi-permanent value shall be 0.7, 0.6 and 0.5 respectively.

7.2.7 The design wind load for railway passenger station with irregular structure should be determined through wind tunnel test. The wind uplift resistance test should be carried out for light metal envelope.

7.3 Materials

7.3.1 The strength grade of concrete shall comply with the following requirements:

1 For the main structure with design durability of 50 years, the strength grade of ordinary reinforced concrete should not be lower than C30 and that of prestressed concrete should not be lower than C40.

2 For the main structure with design durability of 100 years, the strength grade of ordinary reinforced concrete shall not be lower than C35 and that of prestressed concrete shall not be lower than C40.

3 The strength grade of the concrete of track-bearing storey and the structures below shall not be lower than C40.

7.3.2 Reinforcing steel shall comply with the following requirements:

1 HPB300, HRB400 or HRB500 steel bars should be used as the longitudinal load-carrying ordinary bars for track-bearing storey and the structures below, and their properties shall comply with the *Code for Design of Concrete Structures of Railway Bridge and Culvert* (TB 10092); HPB300, HRB400, HRB500, HRBF400 or HRBF500 steel bars should be used for other structures and their properties shall comply with the *Code for Design of Concrete Structures* (GB 50010).

2 Prestressing wires, steel strands, or prestressing deformed bars should be used as the prestressing steel.

7.3.3 The bridge-dedicated low-alloy structural steel of Grade Q235q (D) and Q345q, Q370q or

Q420q (D, E) should be used for track-bearing storey and the structures below; the carbon structural steel of Grade Q235 (B, C and D) and the low-alloy high-tensile structural steel of Grade Q345 (B, C, D and E) should be used for other structures. Other steel types and grades may be selected if reliable evidence is available.

7.4 Special Structure

7.4.1 The bridge-building integrated structure shall comply with the following requirements:

1 The track-bearing storey and the structural members below shall comply with the relevant standards for railway bridges and civil buildings.

2 For the main line of passenger railway with train speed greater than 160 km/h, the main line of freight railway with train speed greater than 120 km/h as well as the railway lines for double-deck container trains or heavy-haul trains, the bridge-building integrated structure should not be employed.

3 Global analysis shall be carried out for the structure based on the spatial layout and the special stress-bearing parts shall be analyzed when necessary. In the case that the bridge-building integrated structure is subjected to various stress conditions during different stages of construction period and service period, structural analysis shall be carried out respectively and the most unfavorable combination of actions shall be determined.

4 The rail-bearing beam shall comply with the following requirements:

1) Analysis of the train load shall be carried out by means of the maximum influence line according to the most unfavorable distribution of dynamic load.

2) For the mixed traffic railway under standard railway live load specified by P. R. C, the vertical deflection limit of the beam shall comply with the *Code for Design of Railway Bridge and Culvert* (TB 10002); for the intercity railway under ZC load, the vertical deflection limit of the beam shall comply with the *Code for Design of Intercity Railway* (TB 10623); for the high-speed railway under ZK load, the vertical deflection limit of the beam shall comply with the *Code for Design of High-speed Railway* (TB 10621).

3) Under the action of lateral sway force of train, centrifugal force of train and wind force, the horizontal deflection limit of the beam shall comply with the *Code for Design of Railway Bridge and Culvert* (TB 10002).

5 Performance-based seismic design shall be carried out for the beam and column members of track-bearing storey of structures in regions with the seismic precautionary intensity of Degree Ⅶ (Site Class Ⅳ), Degree Ⅷ or above.

6 The dynamic response of vehicle-bridge coupling should be analyzed for structures.

7.4.2 The long-span roof structure shall comply with the following requirements:

1 For structural design, the structural scheme shall be determined based on the architectural scheme, manufacture, installation methods, etc.

2 Special analysis and assessment shall be made for the structural design of long-span roof with one of the following structural features: rarely-used structure type, span length greater than 120 m, structure unit length greater than 300 m, or cantilever length greater than 40 m.

3 Type selection and layout of the long-span roof structure and its supporting structure shall comply with the following requirements:

1) The long-span roof structure and its supporting structure shall be designed with effective force transmission paths.

2) The long-span roof structure and its supporting structure shall have proper distribution of

stiffness and bearing capacity and should be arranged uniformly and symmetrically.

3) The structural layout should not result in vulnerable area formed by local weakening or sudden structural change, and for possible vulnerable areas, measures shall be taken to improve its bearing capacity.

4) The supporting structure below the long-span roof shall be so arranged that the long-span roof structure shall not be subjected to great torsions during earthquakes.

4 The displacement and internal force of the structure under the actions of gravity and wind load shall be calculated. In addition, the displacement and internal force under the actions of earthquake, temperature change, bearing seat settlement, construction and installation loads, etc. shall also be calculated according to the specific conditions, so as to meet the requirements of ultimate limit state and serviceability limit state.

7.4.3 The very long concrete structure shall comply with the following requirements:

1 The influence of temperature effect, concrete material shrinkage and other indirect effects on structure stresses shall be considered in the structural calculation and analysis.

2 Prestressing shall be performed or additional steel bars shall be installed for the very long concrete structure.

3 The low-shrinkage concrete should be used or measures such as alternate bay concrete pouring and post-poured strip may be taken for the very long concrete structure.

7.5 Structure of Underground Station

7.5.1 The structure type and construction method of underground station shall be determined through technical and economic comparison based on the engineering conditions at the station site.

7.5.2 The clearance of structure of underground station shall meet the requirements of railway structure gauge; in addition, the influence of construction error, structural deformation and post-construction settlement shall be taken into account and the aerodynamic effect shall be calculated within the range of the track area.

7.5.3 The structure of underground station shall be designed according to the most unfavorable load combination and the anti-floating stability shall be checked based on the anti-floating water level provided by the geological surveyor.

7.5.4 The selection of structure type of underground station shall comply with the following requirements:

1 The structure type shall match the construction method adopted.

2 Reinforced concrete structure or shaped steel (steel tube) - concrete hybrid structure shall be adopted for the underground station constructed by open-cut method or cover-excavation method. The underground diaphragm wall and cast-in-place pile supporting structure should be designed as part of side walls of the main structure to carry load together with the lining wall. The combination of walls may be of superposed or composite type.

3 The structure constructed by mining method shall be designed as composite lining structure; its cross section, supporting and lining parameters shall be determined according to the surrounding rock conditions, functional requirements, construction method, cross section size, environmental protection requirements, etc.

7.5.5 The design of the structure of underground station shall comply with the following requirements:

1 The station should be built on dense, homogeneous and stable soil. When the railway passenger station is located in geologically unfavorable sections such as weak soil, liquefiable soil or fault fracture zone, the influence of such unfavorable conditions on structure shall be analyzed and measures shall be taken accordingly.

2 The structural layout shall be simple, symmetrical and regular.

3 The calculation of structural strength, stiffness and stability shall be carried out in the construction stage and service stage respectively. In addition, crack resistance check shall also be conducted for reinforced concrete structure. When seismic action or other accidental load effects are taken into account, the crack width may not be carried out.

4 The allowable value of the calculated crack width of reinforced concrete structure shall be determined according to the structure type, functional requirements, environment, waterproof measures, etc.

5 In the case of any of the following conditions, the longitudinal strength and deformation of the structure of underground station shall be analyzed:

1) The overburden load varies substantially along the longitudinal direction.

2) The structure directly bears the large local load of buildings (structures).

3) There is a significant difference in the types of foundations or foundation soils along the longitudinal direction.

4) Differential settlement occurs to the structure along the longitudinal direction.

5) Seismic action.

6 In the case of a large interval between the deformation joints, the influence of temperature change and concrete shrinkage on the structure of underground station shall be considered. The section with obvious spatial stress effects should be analyzed as a three-dimensional structure.

7 In the case of prefabricated members, the requirements for production, installation, transportation and construction shall be taken into consideration.

8 The structure of underground station constructed by mining method shall comply with the *Code for Design of Railway Tunnel* (TB 10003).

7.5.6 The details of the structure of underground station shall comply with the following requirements:

1 Deformation joints should not be arranged on the main structure of underground station. When deformation joint is arranged, reliable measures shall be taken to ensure that the settlement difference affecting the train operation safety and normal operation will not occur to the structures on both sides of the deformation joints. The connection between the main structure of underground station and the auxiliary structure such as station entrance/exit passageway shall be provided with deformation joints.

2 The thickness of concrete cover shall comply with the *Code for Durability Design on Concrete Structure of Railway* (TB 10005).

3 The later-built partition wall shall be connected to the main structure reliably. Reinforced concrete structure shall be adopted for the partition wall near the track area.

7.5.7 The waterproofing of underground station structure shall comply with the following requirements:

1 The waterproofing of the structure shall be of Grade Ⅰ as specified in the *Technical Code for Waterproofing of Underground Works* (GB 50108).

2 The waterproof measures shall be determined according to the structure type and construction method, and measures shall be taken to protect the groundwater resources.

3 The thickness of waterproof concrete of the main structure shall not be less than 300 mm.

4 Additional waterproofing layer shall be provided between the retaining structure or primary support and the main structure. The type and laying method of the waterproofing layer shall be determined according to the environmental conditions, structure type, waterproofing grade, construction method, etc.

7.6 Foundation

7.6.1 Foundation shall be designed according to the geological conditions at the construction site, type and load characteristics of superstructure, construction conditions, service requirements, etc., and shall meet the requirements of foundation soil bearing capacity and settlement control.

7.6.2 For the structure foundation in areas with goaf, unfavorable geological conditions or unfavorable earthquake-resistant conditions, etc., reliable measures shall be taken to prevent instability of foundation soil.

7.6.3 The interaction among foundation soil, foundation, and superstructure shall be considered in the foundation design.

7.6.4 For the foundation shared between an independent bridge separating from the railway passenger station structure and its adjacent structures, the integral foundation settlement and the differential settlement shall be calculated and analyzed, so as to meet deformation control requirements for the purpose of ensuring safe and comfortable train operation.

7.6.5 The foundation of bridge-building integrated structure shall comply with the following requirements:

1 The foundation grade shall not be lower than Grade B.

2 The foundation deformation shall comply with the relevant standards for railway bridges and civil buildings.

3 When pile foundation is adopted, testing of foundation piles shall be carried out according to the requirements of the *Technical Code for Testing of Building Foundation Piles* (JGJ 106).

7.7 Deep Foundation Pit

7.7.1 The deep foundation pit of railway passenger station shall comply with the following requirements:

1 When the railway passenger station is combined with urban rail transit and municipal works, the design of foundation pit shall be performed in a coordinated manner, and should take into consideration the structure and the foundation.

2 The safety grade and technical criteria of foundation pit shall be determined according to the engineering geological and hydrogeological conditions, overall project planning, transition scheme for existing line, surrounding environment, etc.

3 The supporting structure shall meet the calculation and checking requirements for ultimate limit state and serviceability limit state.

4 Deformation control indexes for foundation pit shall be determined according to the requirements for the normal service of supporting structure and the protection of surroundings.

5 Specialized detailed design for foundation pit dewatering and monitoring shall be carried out.

7.7.2 The safety grade of foundation pit shall comply with the following requirements:

1 The safety grade of foundation pit shall be determined according to the excavation depth, influence of supporting structure failure on the surroundings, positional relationship with the railway

operational line, etc.

2 The safety grade of foundation pit shall comply with relevant national and local standards. When the distance from the outer edge of the existing railway earthworks to the outer excavation sideline of foundation pit is less than twice the foundation pit depth, the safety grade of the foundation pit shall be Grade I.

7.7.3 The design of foundation pit shall comply with the following requirements:

1 The horizontal loads acting on the supporting structure shall not only include the water and earth pressure, but also include the load of existing railway earthworks, the horizontal load caused by train and track load, as well as the horizontal load acting on the later-stage foundation pit caused by overloading of the early-stage station building during staged construction of station building works.

2 The load of existing railway earthworks may be applied to the supporting structure in the form of ground overload.

3 The train and track load may be converted into soil column load before being applied to the supporting structure.

7.7.4 The excavation and backfilling of foundation pit shall comply with the following requirements:

1 For the excavation of foundation pit, the time-space effect shall be fully utilized and the construction method of supporting prior to excavation shall be adopted so as to ensure balance and symmetry of excavation. In addition, over-excavation is prohibited.

2 Earthwork excavated from the foundation pit should be utilized for backfilling of side walls of the foundation pit, while the technical indexes of backfill shall meet the technical requirements for earthworks, buildings and other works.

7.7.5 Risk control of foundation pit shall comply with the following requirements:

1 For the design of foundation pit, targeted risk control scheme and emergency plan shall be developed according to the engineering geological and hydrogeological conditions, surrounding environment and other characteristics.

2 If the foundation pit is close to the railway operational line, a specialized monitoring scheme shall be developed for real-time tracking and monitoring of earthworks and track deformation.

3 Targeted risk prevention and control measures shall be taken according to the structure type, foundation type and protection requirements of the adjacent existing buildings (structures).

7.8 Canopy and Platform Wall

7.8.1 The structural type of the canopy with column on platform shall be selected according to the architectural form, environment and other factors. For the canopies located in high wind pressure regions, coastal regions, acid rain regions, etc., reinforced concrete structures should be employed.

7.8.2 For the railway passenger station designed with track-bearing storey structure, the settlement difference between the platform wall on the track-bearing storey and the platform wall outside shall not affect the normal operation, and measures such as ground treatment, setting of transition sections and temporary loose-fill paving may be taken to control the settlement difference; for platform walls located in areas with soft soil or other unfavorable geological conditions, ground treatment shall be carried out in combination with the earthwork design.

7.8.3 The foundation of the canopy with columns on platform shall be coordinated with the platform beam, retaining wall for earthworks, underpass and other structures; the canopy of the corridor between the station building and the main platform shall be coordinated with the structural layout of station building and retaining wall for earthworks.

7.8.4 When a structural system is shared between the canopy and OCS or sign boards of the passenger guidance system etc., coordinated design shall be performed in respect of the structural layout, load, structural details, etc.

7.9 Others

7.9.1 The buildings near the railway line must meet the requirements of railway structure gauge with consideration to the construction error and structural deformation.

7.9.2 In the case of upgrading of existing railway passenger station, reliability appraisal shall be carried out for the existing structures in areas with functional change and the design scheme for upgrading shall be determined according to the appraisal conclusion, the subsequent service requirements and service life.

7.9.3 In the cases of bridge-building integrated station, the comfort levels of station platform storey and elevated waiting area storey and the deck of passenger overpass shall meet the requirements of relevant standards. When crossing the main line, the influences of vibration and wind pressure caused by train operation on the structural comfort level shall be considered.

7.9.4 In the case of construction and upgrading of railway passenger stations, the factors of the existing railway line shall be considered and the structural design shall satisfy various boundary conditions, with consideration to the different stress conditions at the construction stage and the service stage.

7.9.5 In the case of steel structure, the materials shall be selected according to the atmospheric conditions of the project and anti-corrosion design shall be carried out according to the structure types, the anti-corrosion design service life and construction conditions, etc. The construction environment of welding, painting and other processes shall comply with the technological requirements. The open-air steel structures shall be made of weather-resistant steel. The derusting grade shall not be lower than Sa2 $\frac{1}{2}$ and the weather-resistant paint shall be adopted for the anti-corrosion coating. The anti-corrosion design service life shall not be shorter than 15 years.

7.9.6 Reasonable construction processes shall be developed for the construction of the connection point between station building and other relevant railway works such as station and yard, earthworks and bridge.

7.9.7 Specialized construction technology scheme shall be developed for the hoisting of large-scale steel structures. The complex and important steel structures shall be pre-assembled. The removal sequence and procedures of temporary support structure of large-span steel structures shall be determined through calculation, and analysis and assessment shall be carried out when necessary.

7.9.8 Before the construction of mass concrete and concrete of very long structures, specialized construction scheme shall be developed and the mixing proportion of concrete shall be determined through trial mixing.

8 Heating, Ventilation and Air Conditioning

8.1 General Requirements

8.1.1 The scheme for heating, ventilation and air conditioning shall be determined according to the scale of the passenger station building, surrounding environment, local energy conditions, national policies on energy conservation, emission reduction, environmental protection, etc.

8.1.2 The air conditioning system in the equipment area of the station building shall be set separately. The air conditioning system in the public area and that in the office area should be separated.

8.1.3 The cold and heat sources of the heating and air-conditioning systems, as well as the energy-consuming equipment such as the fans and water pumps of the transmission and distribution system, shall be of National Grade I energy efficiency.

8.2 Indoor and Outdoor Atmospheric Design Parameters

8.2.1 The indoor design temperature for heating in the main rooms of the station building shall comply with Table 8.2.1.

Table 8.2.1 Indoor Design Temperature for Heating in Main Rooms of Station Building

Room	Indoor Design Temperature for Heating (℃)
Entry concourse	12～14
Ticket hall, baggage deposit hall and reception hall	14～16
Waiting area (hall/room)	18
Business-class lounge	20
Room for commercial use	18
Office and ticket office	18
Public toilet	14～16
Bill room	10
Baggage storehouse	5
Passenger underpass	No heating
Equipment room	According to the technical requirements

Notes: 1 When the low-temperature floor-radiation heating system is adopted, the indoor design temperature for heating shall be 2℃ lower than that specified in the table.

2 When the exit concourse is arranged indoors and connected to the heating zone, the heating temperature shall be the same as that of the entry concourse and heating is not required when the exit concourse is arranged outdoors.

8.2.2 The indoor design temperature for air conditioning and the relative humidity in each main room of the station building shall comply with Table 8.2.2.

Table 8.2.2 Indoor Design Temperature for Air Conditioning and Relative Humidity in Each Main Room of Station Building

Room	Indoor Design Temperature for Air Conditioning in Summer (℃)	Relative Humidity (%)
Entry concourse	28～30	40～70
Ticket hall and waiting area (hall/room)	26～28	40～70

Table 8.2.2(continued)

Room	Indoor Design Temperature for Air Conditioning in Summer (℃)	Relative Humidity (%)
Business-class lounge	24～26	40～60
room for commercial use	26～28	40～60
Office and ticket office	26	40～60
Public toilet	27～28	40～80
Equipment room	According to the technical requirements	

Notes: 1 When the exit concourse is arranged indoors and connected to the air conditioning zone, the temperature for air conditioning shall be the same as that of the entry concourse and air conditioning is not required when the exit concourse is arranged outdoors.

2 The indoor design temperature for air conditioning in winter may be determined based on the indoor design temperature for heating.

8.2.3 When the ventilation system is adopted in the station building, the indoor design air temperature in the public area in summer should not be higher than the outdoor design temperature for air ventilation by more than 5℃ and shall not be higher than 30℃.

8.2.4 The fresh air volume of air conditioning system in each main room of the station building shall comply with Table 8.2.4.

Table 8.2.4 Fresh Air Volume of Air Conditioning System in Each Main Room of Station Building

Room	Minimum Fresh Air Volume [m^3/(h·person)]
Ordinary waiting area (aboveground station and elevated station)	10.0
Ordinary waiting area (underground station)	12.6
First-class lounge (independent)	20.0
Business-class lounge	30.0
Ticket hall	10.0
Room for commercial use	20.0
Ticket office	30.0

Notes: 1 When the first-class lounge is arranged in the ordinary waiting area, it shall be considered as ordinary waiting area in calculation.

2 The fresh air volume at underground station shall meet the requirements of Table 8.2.4 and shall not be less than 10% of the total air volume of the air conditioning system.

3 Fresh air system is not required in the concourse.

8.2.5 The outdoor atmospheric design parameters shall comply with *Design Code for Heating Ventilation and Air Conditioning of Civil Buildings* (GB 50736).

8.3 Heating

8.3.1 Heating in the station building in winter shall comply with the following requirements:

1 In regions where the number of days with average temperature being lower than or equal to 5℃ for years is more than or equal to 90 d, heating facilities shall be provided and central heating should be adopted in the station building.

2 Local heating may be adopted for the equipment room and management room with heating requirements.

8.3.2 The calculation of heating load in the station building shall consider the factors of the porches and passenger flow line.

8.3.3 The low-temperature hot-water floor-radiant heating should be adopted for the ticket hall, concourse and waiting area (hall/room) of station building in severe cold and cold regions.

8.3.4 Hot air curtains shall be provided at the main entrances and exits of the station building in severe cold regions and should be provided at the main entrances and exits of the station building in cold regions; air curtains should be provided at the main entrances and exits of the station building in hot-summer warm-winter regions and in hot-summer cold-winter regions.

8.4 Ventilation

8.4.1 Natural ventilation should be adopted at the waiting area and ticket hall of the station building. When natural ventilation cannot meet the requirements, auxiliary mechanical ventilation shall be adopted. Mechanical ventilation shall be adopted at the public toilet. The ventilation rate should comply with Table 8.4.1.

Table 8.4.1 Ventilation Rate for Mechanical Ventilation in Station Building

Room	Ventilation Rate (times/h)
Waiting area (hall/room) and ticket hall	2～3
Public Toilets (mechanical ventilation)	15～20

8.4.2 Mechanical ventilation shall be provided in the offices without windows facing outdoors.

8.4.3 In the station building, automatic temperature control fans shall be provided in electrical rooms provided with mechanical ventilation for residual heat removal.

8.4.4 Independent ventilation system shall be provided in foodstuff-operation room in the public area of the station building.

8.4.5 Independent mechanical ventilation system shall be provided in toilets in the station building and the exhaust air shall be directly discharged outdoors.

8.5 Air Conditioning

8.5.1 In hot-summer cold-winter regions and in hot-summer warm-winter regions, air conditioning system shall be provided in the very large, large and medium station buildings as well as frontier (port) station buildings, and should be provided in the small station buildings; in other regions, air conditioning system may be provided in the station building. Independent air conditioning system should be provided in the business-class lounge and ticket office.

8.5.2 The air conditioning system shall be provided in the equipment room of station building according to the technical requirements.

8.5.3 In the areas with high radiant heat in the station building, the floor-radiant heating system may be adopted for cooling in summer, and dewing-proof measures shall be taken.

8.5.4 Stratified air conditioning should be adopted for high and large space such as the waiting area (hall/room), concourse and ticket hall.

8.5.5 Air-purifying device should be provided in the public area of the railway passenger station provided with all-air air-conditioning system.

8.6 Others

8.6.1 Proper operation strategies should be prepared for the heating, ventilation and air conditioning system.

8.6.2 For the heating and air conditioning system, measures shall be taken to reduce energy consumption when partial cooling and heating load is used or the system is used locally.

8. 6. 3 Measures should be taken to control fresh air demand in the waiting area (hall/room) and ticket hall of the station building.

8. 6. 4 Energy measuring instruments shall be provided for the heating, ventilation and air conditioning system and shall comply with the *General Principle for Equipping and Managing of the Measuring Instrument of Energy in Organization of Energy Using* (GB 17167) and *Code for Design of Energy-saving of Railway Engineering* (TB 10016).

8. 6. 5 Maintenance facilities shall be provided for the heating, ventilation and air conditioning system in the station building.

8. 6. 6 The heating, ventilation and air conditioning equipment and pipeline below the elevated station building shall not be arranged above the railway line; otherwise, protective measures shall be taken.

8. 6. 7 The form and layout of the terminal devices for heating, ventilation and air conditioning system shall be in harmony with the building decoration.

9 Water Supply and Drainage

9.1 General Requirements

9.1.1 The water supply system of railway passenger station shall be designed according to the production, living, firefighting water requirements and the requirements of water quality, water pressure and water quantity.

9.1.2 The drainage system of railway passenger station shall be designed according to the separation of domestic sewage and rainwater.

9.1.3 The water-saving sanitary wares and accessories shall be adopted in the station building.

9.1.4 The design of drinkable water facilities at passenger station shall comply with relevant standards.

9.1.5 The design of water supply and drainage works of railway passenger station shall comply with not only this Code, but also *Code for Design of Water Supply and Drainage for Railway* (TB 10010).

9.1.6 The water supply and drainage equipment and pipeline below the elevated station building shall not be arranged above the railway line; if such arrangement is required, safety protection measures shall be taken. The thermal insulation and anti-freezing measures shall be taken for the water supply and drainage pipeline below the elevated station building in severe cold and cold regions.

9.2 Water Supply

9.2.1 Urban water supply sources should be preferred for railway passenger station, the pressurization, water storage and water treatment facilities shall be provided according to the water quality, water consumption, water pressure and guarantee rate of water supply.

9.2.2 Circulating water and reused water should be adopted for the railway passenger station according to the nature of water use.

9.2.3 The water consumption by passengers in the station building shall be calculated according to Formula (9.2.3). The water consumption quota for passengers and the coefficient of water consumption non-uniformity shall comply with those specified in Table 9.2.3. The hourly variation coefficient within the water supply time should be 2.0～3.0 for passenger station building of mixed traffic railway and should be 2.5～3.0 for passenger station building of intercity railway and high-speed railway.

$$Q=\alpha \cdot H \cdot q_g \times 10^{-3} \qquad (9.2.3)$$

Where

Q——water consumed by passengers in the station building (m^3/d);

α——coefficient of water consumption non-uniformity;

H——maximum number of passengers gathered in railway station building;

q_g——water consumption quota for passengers [L/(d · person)].

Table 9.2.3 **Water Consumption Quota for Passengers and Coefficient of Water Consumption Non-uniformity**

Classification of Passenger Station Building	Water Consumption Quota (maximum daily)[L/(d • person)]	Coefficient of Water Consumption Non-uniformity
Passenger station building of mixed traffic railway	15～20	2.0～3.0
Passenger station building of intercity railway and high-speed railway	3～4	1.0～2.0

9.2.4 Hot water shall be supplied in washrooms in the public area of very large and large station buildings in severe cold and cold regions.

9.2.5 The passenger station building of mixed traffic railway should be provided with drinking water equipment as per an index of 1.0 L/(d • person)～2.0 L/(d • person); the passenger station building of intercity railway and high-speed railway should be provided with drinking water equipment as per an index of 0.2 L/(d • person)～0.4 L/(d • person). The coefficient of hourly variation within the drinking water supply time should be 1.0.

9.2.6 The water supply facilities used for cleaning should be provided on both ends of the intermediate platform of the railway passenger station.

9.3 Drainage

9.3.1 The sewage from the railway passenger station should be discharged into the urban drainage pipe network. For regions where the urban drainage pipe network is not accessible, the quality of sewage discharged shall comply with the requirements of the national or local discharge standards.

9.3.2 The reuse of reclaimed water and the comprehensive utilization of rainwater shall comply with relevant standards.

9.3.3 Overflow installations shall be provided for the siphonic roof-top rainwater drainage system and the siphon mouths of the same system shall be of the same elevation.

9.3.4 The diameter of the sewage (not including the sewage produced in production) pipes in the public area of passenger station building shall be increased by one level compared with the calculated pipe diameter.

10 Electric Power and Lighting

10.1 General Requirements

10.1.1 The power supply capacity of railway passenger station shall be determined by giving overall consideration to the electric equipment for communication, signaling, passenger service, automatic fire alarm, equipment monitoring, ventilation and air conditioning, lifts and escalators, as well as the commercial facilities for passenger service, etc.

10.1.2 The safe and reliable electric power equipment and lamps with economic applicability and high energy efficiency shall be selected in the electric power and lighting design of railway passenger station.

10.1.3 Electrical and mechanical control system may be provided at railway passenger station as required.

10.1.4 The electric power and lighting design shall be coordinated with the architecture, structure and other disciplines as well as the related municipal lighting design.

10.1.5 In addition to this Code, the electric power and lighting design of railway passenger station shall also comply with other relevant current standards.

10.2 Power Supply and Distribution

10.2.1 Substations for medium or larger passenger stations should be located inside the station building.

10.2.2 The Class I load for the railway passenger station shall be supplied by dual power source. The electrical load rating and power supply shall comply with the *Code for Design of Railway Electric Power* (TB 10008).

10.2.3 Diesel generator sets shall be provided at very large passenger stations and should be provided at important large passenger stations and medium or larger underground passenger stations.

10.2.4 Emergency power supply (EPS) units should be adopted for the emergency lighting and evacuation signs at railway passenger station, or the batteries and float charging devices that accompany the lighting fixtures should be used as the backup power source.

10.2.5 When the railway passenger station is provided with photovoltaic power generation system, cogeneration system or other distributed power generation systems, the equipment configuration, main wiring form and operating mode shall comply with requirements of grid connection, etc.

10.2.6 The low-voltage distribution trunk line system of railway passenger station shall comply with the following requirements:

1 The electrical loads in the same area and of the same category should share the same distribution trunk line.

2 The advertising and commercial facilities as well as other loads requiring separate accounting of cost should be supplied by the circuits separately from the operating load of the passenger station.

3 Dedicated power distribution circuits should be adopted for lighting in the public area.

4 Dedicated power distribution circuits shall be adopted for emergency lighting and firefighting load. For medium or larger passenger stations, the dedicated power distribution circuits may be separated from the regional main distribution equipment according to the fire zoning and building layout.

10.3 Lighting

10.3.1 The types of lighting for station building shall comply with the following requirements:

1 Zoned general lighting may be adopted in the public area according to the functional requirements.

2 Accent lighting should be adopted at the ticket counter, security inspection area, service counter, wash stand and other areas.

3 General lighting shall be adopted in the office area and equipment area.

10.3.2 The categories of lighting for station building shall comply with the following requirements:

1 General lighting shall be provided in the public area, office area and equipment area and lighting arrangement should facilitate group control.

2 Emergency lighting shall be provided in the public area and the signaling, firefighting and alarm equipment rooms, as well as other important rooms directly related to train operation.

3 Obstruction lighting shall be provided on the station building affecting aviation safety according to relevant regulations.

4 Landscape lighting may be provided as required.

10.3.3 In the areas where the mounting height of luminaire is 6 m or below, the three-band T5 or T8 fluorescent lamps, compact fluorescent lamps or light emitting diode (LED) lamps should be adopted. The metal halide lamps or light emitting diode (LED) lamps should be adopted in the areas where the mounting height of luminaire exceeds 6 m.

10.3.4 Direct lighting should be adopted in high and large space in station building and lamps should be arranged in groups in a centralized manner.

10.3.5 In the station building, the lighting design of the public area where the open-type commercial room is located shall be performed by giving overall consideration to zoning control, luminance control and other aspects.

10.3.6 When fluorescent lamps and LED lamps are used for general lighting, part of them may be used for emergency lighting; when metal halide lamps are used for general lighting, the halogen tungsten lamps or LED lamps may be used for emergency lighting.

10.3.7 The layout of lighting lamps in the platform area shall meet the requirements for installation, maintenance and the safety distance of live parts of OCS.

10.3.8 The lighting lamps shall be installed on the structure with sufficient strength. Lamps and lighting sources in the public area of the station building shall be fixed firmly and anti-falling measures shall be taken.

10.3.9 The position and installation of lighting lamps and related equipment shall meet the requirements for maintenance in the future and dedicated maintenance equipment may be provided when necessary.

10.3.10 The design scheme of landscape lighting for railway passenger station shall be determined based on the layout of municipal works, architectural features and local culture, and the landscape lighting shall be in harmony with the surrounding buildings.

11 Passenger Transport Service Information System

11.1 General Requirements

11.1.1 The railway passenger station shall be provided with passenger transport service information systems including passenger service information system, ticketing system, baggage information system, access control system, etc.

11.1.2 The passenger transport service information system shall match the scale and operation management mode of the railway passenger station.

11.1.3 The terminal of the passenger transport service information system shall be arranged in coordination with the architecture, structure and decoration of the station building on the premise of meeting the requirements of passenger service and operation management.

11.1.4 The design of passenger service information system shall reserve conditions for future development.

11.2 Passenger Service Information System

11.2.1 The passenger service information system may be comprised of subsystems such as integrated management platform, public address system (PA), passenger information display system, video surveillance system, clock, security inspection equipment, information enquiry system, intrusion alarm system, help system, etc.

11.2.2 The integrated management platform shall exercise centralized control of public address system, passenger information display system, enquiry system, etc.

11.2.3 The public address system shall comply with the following requirements:

1 The public address system shall provide public announcements for passengers during the course of ticket purchasing, entering, waiting, boarding and exiting, and shall provide service announcements for passenger service personnel.

2 Loudspeakers should be installed at the station square, ticket hall, entry concourse, waiting area (hall/room), platform, exit concourse, working area, commercial area, etc.

3 Loudspeakers shall be arranged according to the architecture, structure and decoration of the station building as well as their electrical performance indexes, and shall comply with the *Technical Code for Public Address System Engineering* (GB 50526) and Article 5.8.5 of this Code.

4 The appearance, color, structure and installation of loudspeakers shall be in harmony with the environment of the railway passenger station.

5 When the public address system is also used for emergency public address in case of fire, it shall comply with *Code for Design of Automatic Fire Alarm System* (GB 50116).

11.2.4 The passenger information display system shall comply with the following requirements:

1 The passenger information display system shall provide guidance and information for passengers and provide service information for passenger service personnel.

2 Displays should be set at ticketing hall, entry concourse, waiting area (hall/room), platform, and exit concourse according to the entering and exiting flow lines.

3 The electronic displays should be in coordination with the clocks, static signs, etc.

4 The specifications and installation of the electronic displays shall be determined according to the passenger service requirements and building layout, and functionality, esthetics and economic efficiency shall also be considered.

11.2.5 The video surveillance system shall comply with the following requirements:

1 The video surveillance system shall be used to monitor the area of passenger service and to provide real-time video for the staff at station.

2 Surveillance cameras shall be set at the station square, ticketing hall, entry gate, ticketing area, entry concourse, exit concourse, waiting area (hall/room), platform, staircase and escalator, lift, passage and office area.

3 High-definition digital surveillance cameras should be adopted.

11.2.6 Slave clocks should be provided at the ticket hall, entry concourse, waiting area (hall/room), platform, exit concourse, etc.

11.2.7 The security inspection equipment, metal detector door, portable metal detectors, explosion-proof tanks, explosion-proof blankets, etc. shall be arranged at stations. Explosives detectors or liquid detectors may be provided as required.

11.2.8 Enquiry terminals should be arranged at the ticket hall, waiting area (hall/room), concourse, information center, passenger service desk, etc.

11.2.9 The detectors of the intrusion alarm system shall be arranged at the bill room, income office, etc.; the emergency alarm button shall be arranged at the manual ticket counter, ticket office passage, excess fare room, etc.

11.2.10 The on-site help-seeking equipment should be arranged at the ticketing hall, entry concourse, waiting area (hall/room), platform, exit concourse, etc.

11.3 Ticketing System

11.3.1 The ticketing system shall provide the functions of ticket purchasing, ticket fetching, ticket refunding, excess fare paying and ticket checking, and should provide the functions of ticket loss reporting and replacement, ticket change and rescheduling. In addition, it may provide the functions of identity verification and ticket checking as required.

11.3.2 The ticketing system should be provided with servers and management terminals. In addition, booking office machine, ticket vending machine, ticket fetching machine, automatic ticket gate, ticket machine for excess fare, identity verification equipment, etc. may be provided as required.

11.3.3 The number of booking office machines and ticket vending machines shall be determined according to the number of passengers buying tickets at the station, processing speed of the terminal, working hours of the terminal, etc. The number of booking office machines and ticket vending machines should not be less than two respectively at each station.

11.3.4 The number of ticket fetching machines shall be determined according to the number of passengers fetching tickets at the station, the terminal processing speed and other parameters.

11.3.5 The number of automatic ticket gates shall be determined according to the number of check-in passengers, terminal processing speed and check-in time, etc., and shall not be less than two at each group of check-in gates.

11.4 Others

11.4.1 The baggage management information system at the station should be provided with baggage management server, computer, bill printer, label printer, network equipment, etc.

11.4.2 The baggage service information system shall comply with the following requirements:

1 The baggage service information system shall be provided with the information display subsystem, public address subsystem, video surveillance subsystem, security inspection subsystem, etc.

2 The PA host, control and transmission equipment may be integrated with the passenger service information system according to the operation management mode.

3 The video storage and management equipment should be integrated with the passenger service information system according to the operation management mode.

11.4.3 Access control system shall be installed at the information room, ticket office, bill room, income office, excess fare room, etc., and should be installed at the room of intermediate distribution frame. The access control system may be installed at the important offices of the station as required.

12 Accessible Facilities

12.0.1 The accessible flow line of railway passenger station shall be continuous and complete and shall be connected with the municipal accessible traffic facilities.

12.0.2 The accessible facilities at railway passenger station shall cover the terrace for station building, public area and facilities for passenger transport, etc., and shall meet the demands of passengers with reduced mobility for ticket purchasing, waiting, check-in, check-out, and baggage deposit.

12.0.3 When there is a height difference between the terrace for station building and the station square, curb ramps or wheelchair ramps shall be arranged. Accessible lift shall be provided if it is difficult to arrange wheelchair ramps. The lifting platform may be provided if it is difficult to arrange accessible lifts. The curb ramps and wheelchair ramps shall be close to the main entrances and exits of the facilities and buildings related to passengers with reduced mobility.

12.0.4 The accessible facilities in the concourse shall comply with the following requirements:

1 The entrance and exit of concourse shall be accessible.

2 When there is a height difference between the entry concourse and the waiting area (hall/room) and between the concourse and the ground level, the wheelchair ramps, accessible lifts, lifting platform or other lifting facilities shall be provided.

3 Where the ground level of the exit concourse varies, the wheelchair ramps, accessible lifts, lifting platform or other lifting facilities shall be provided.

4 At least one low window shall be provided in the identity verification area and the clear width of the ticket-checking passage shall not be less than 0.9 m.

12.0.5 The entrance and exit of waiting area (hall/room) shall be accessible and the wheelchair areas shall comply with the following requirements:

1 The wheelchair areas should be close to the entry gates and the accessible lifting facilities, and may be centralized in different zones.

2 The floor area for each wheelchair shall not be less than 1.10 m×0.80 m.

3 The passage leading to the wheelchair areas shall comply with Article 12.0.7 of this Code.

4 Signs indicating accessibility shall be provided on the ground of the accessible areas.

12.0.6 The accessible facilities in the ticket hall and the baggage office shall comply with the following requirements:

1 The entrance and exit of ticket hall and baggage office shall be accessible.

2 At least one low ticket counter shall be provided.

12.0.7 The passages, corridors, halls (rooms) and cross-track facilities serving passengers with reduced mobility shall comply with the requirements for accessibility. The width of accessible passage shall not be less than 1.50 m, and should not be less than 1.80 m at very large and large passenger stations. The clear width of ticket gates serving the passengers with reduced mobility shall not be less than 0.90 m and the ground within 1.80 m of the railing (two sides) at the ticket gates shall be flat.

12.0.8 Stairs and steps serving the passengers shall be accessible and warning tactile paving with a width of 300 mm~600 mm shall be arranged at a distance of 250 mm~500 mm from the starting point and ending point of the stairs and steps. The length of the warning tactile paving shall be the same as the width of the stairs or steps.

12.0.9 The ramps serving for the passengers with reduced mobility other than the curb ramps and wheelchair ramps shall comply with the following requirements:

1 The ramp gradient shall not be greater than 1/12.

2 The ramp surface shall be smooth and skid-proof.

3 The clear width of the ramp shall not be less than 2.0 m.

4 An intermediate platform with a length not less than 2.0 m shall be provided every time the ramp raises 1.5 m.

5 Handrails shall be provided on both sides of the ramp and shall comply with the *Code for Accessibility Design* (GB 50763). Safety barriers should be arranged at the bottom of the handrails.

6 Warning tactile paving with a width of 300 mm~600 mm shall be arranged at a distance of 250 mm~500 mm from the starting point and ending point of each ramp and the length of the sidewalks shall be same as the width of the ramp.

12.0.10 Ramps or accessible lifting facilities shall be provided to connect the cross-line facilities and each platform serving for passengers with reduced mobility. The accessible lifting facilities shall comply with the following requirements:

1 The accessible lifts leading to the platform shall be provided at very large and large passenger stations.

2 In case it is difficult to arrange ramps at medium passenger stations, the accessible lifts leading to the platform shall be provided or lift shaft shall be reserved. When the lift shaft is reserved, lifting facilities such as accessible lifting platform or stairlift shall be provided.

3 In case it is difficult to arrange ramps at small passenger stations, lifting facilities such as accessible lifting platform or stairlift shall be provided.

4 In case there is difficulty in arranging ramps or accessible lifts at upgraded passenger stations, lifting facilities such as accessible lifting platform or stairlift shall be provided.

5 Warning tactile paving with a width of 300 mm~600 mm shall be arranged at a distance of 250 mm~500 mm from the entrance of the accessible lifts and the length of the sidewalks shall be the same as the width of the lift entrance.

6 The location of the call and control buttons of the accessible lifting platform or stairlift shall facilitate the use of passengers with reduced mobility.

12.0.11 The accessibility design of public toilets for passengers shall comply with the following requirements:

1 Dedicated accessible toilets shall be designed at medium or larger passenger stations.

2 Dedicated accessible toilets should be designed at small passenger stations; if difficulty is present, the accessible toilet cubicle shall be provided in the public toilets and shall comply with the *Code for Accessibility Design* (GB 50763).

3 For the railway passenger station with unisex toilet, the unisex toilet shall also serve as the dedicated accessible toilet.

12.0.12 The accessibility design of platform at the passenger station shall comply with the following requirements:

1 Warning tactile paving with a width of 600 mm shall be arranged at the inner side of the warning line on the platform and the length of the warning tactile paving should be the same as that of the warning line. Directional tactile paving shall be provided connecting the warning tactile paving at the inner side of the warning line on the platform with the warning tactile paving at the entrances and exits of stairs, ramps and accessible lifts on the platform that are connected to the cross-line facilities dispatching passengers.

2 The top surface of well covers and gratings shall be flush with the ground and the width of holes on the grating shall not be more than 10 mm.

3 The height from the ground to the lower edge of the objects or signs fixed on the walls and columns shall not be less than 2.0 m. Protective facilities shall be provided in the triangular area with a clear height less than 2.0 m under the escalators and stairs where passengers can enter. In addition, warning tactile paving shall be arranged outside the protective facilities.

4 The British Pendulum Number (BPN) of the tactile paving on the platform shall not be less than 80.

12.0.13 Escalators and the level crossing within the range of the station and yard must not be used as accessible passage. Warning tactile paving with a width of 300 mm~600 mm shall be arranged at a distance of 250 mm~500 mm from the upper and lower supporting points of the escalators. The length of the sidewalks shall be the same as the width of the escalators and the sidewalks must not be connected with the directional tactile paving.

Words Used for Different Degrees of Strictness

In order to mark the differences in executing the requirements in this Code, words used for different degrees of strictness are explained as follows:

(1) Words denoting a very strict or mandatory requirement:

"Must" is used for affirmation; "must not" is used for negation.

(2) Words denoting a strict requirement under normal conditions:

"Shall" is used for affirmation; "shall not" is used for negation.

(3) Words denoting a permission of slight choice or an indication of the most suitable choice when conditions permit:

"Should" is used for affirmation; "should not" is used for negation.

(4) "May" is used to express the option available, sometimes with the conditional permit.

Normative References

Structure Gauge for Standard Gauge Railways (GB 146. 2)

Safety Signs and Guideline for the Use (GB 2894)

Graduations and Test Methods of Air Permeability, Watertightness, Wind Load Resistance Performance for Building External Windows and Doors (GB/T 7106)

Public Information Graphical Symbols for Use on Sign (GB/T 10001)

Guidance System for Public Information-Setting Principles and Requirements (GB/T 15566)

Specification and Test Methods for Road Traffic Markings (GB/T 16311)

General Principle for Equipping and Managing of the Measuring Instrument of Energy in Organization of Energy Using (GB 17167)

Guidance System for Public Information-Design Principles and Requirements for Guidance Elements (GB/T 20501)

Code for Design of Concrete Structures (GB 50010)

Code for Fire Protection Design of Buildings (GB 50016)

Technical Code for Waterproofing of Underground Works (GB 50108)

Code for Design of Automatic Fire Alarm System (GB 50116)

Code for Design of Metro (GB 50157)

Code for Thermal Design of Civil Building (GB 50176)

Design Standard for Energy Efficiency of Public Buildings (GB 50189)

Code for Fire Prevention in Design of Interior Decoration of Buildings (GB 50222)

Calculation Code for Construction Area of Building (GB/T 50353)

Technical Code for Public Address System Engineering (GB 50526)

Design Code for Heating Ventilation and Air Conditioning of Civil Buildings (GB 50736)

Code for Accessibility Design (GB 50763)

Code for Design of Railway Bridge and Culvert (TB 10002)

Code for Design of Railway Tunnel (TB 10003)

Code for Durability Design on Concrete Structure of Railway (TB 10005)

Code for Design of Railway Electric Power (TB 10008)

Code for Design of Water Supply and Drainage for Railway (TB 10010)

Code for Design of Energy-saving of Railway Engineering (TB 10016)

Code for Design of Emergency Evacuation and Rescue Works for Railway Tunnel (TB 10020)

Code for Design on Fire Prevention of Railway Engineering (TB 10063)

Code for Design of Concrete Structures of Railway Bridge and Culvert (TB 10092)

Code for Design of Railway Station and Terminal (TB 10099)

Evaluation Standard for Green Railway Passenger Stations (TB/T 10429)

Code for Design of High-speed Railway (TB 10621)

Code for Design of Intercity Railway (TB 10623)

Code for Design of Office Building (JGJ 67)
Technical Code for Testing of Building Foundation Piles (JGJ 106)
Technical Specification for Skylight and Metal Roof (JGJ 255)
Technical Specification for Slip Resistance of Building Floor (JGJ/T 331)
Standard for Design of Urban Public Toilets (CJJ 14)